Birds
for all Seasons

Birds for all Seasons

JEFFERY BOSWALL

BBC PUBLICATIONS

Published by BBC Publications

A division of BBC Enterprises Ltd

35 Marylebone High Street, London W1M 4AA

First published 1986

ISBN 0 563 20453 2

Set in 12 on 14 point Garamond
by Ace Filmsetting Ltd, Frome, Somerset
Printed in England by W. S. Cowell Ltd, Ipswich, Suffolk

Contents

'O blessed bird! the earth we pace
Again appears to be
An unsubstantial faery place
That is fit home for thee!'

William Wordsworth

Introduction

One problem all of the earth's organisms share is that they have had to come to terms with the fact that time is divided into years, self-repeating years that are themselves divided into seasons. And the seasons exist and the years have tangible points at which they begin again for one main reason: our planet's axis tilts at an angle – $66\frac{1}{2}$ degrees – to the plane of its rotation around its parent star.

If the earth sat at 90 degrees – and there is no reason outside of pure accident that it doesn't – there would still be life on it. The poles would still be colder, the equator would still be hotter and the days would still be twenty-four hours long. But there would be no seasons, and no way of noting the passage of a year. Life would be unrecognisably different in form and behaviour. Just *how* different would make a good theme for a science-fiction story, but just how the actual $66\frac{1}{2}$ degrees has influenced the evolution of the life that is actually here can be judged by the sheer scale and totality of the climatic conditions the tilt has caused.

As our yearly round proceeds, the angle of the planet's attitude to the star steadily changes, and with it the angle at which the one receives its heat and light from the other. This means that any given spot on earth outside a narrow equatorial belt receives steadily less and then steadily more radiation, as if it were being moved north and south across the planet's surface, and life has had to adjust to what amounts to an involuntary migration.

Also, the fluctuating temperature has physical influence on the planet's fluid media – its air and water – and makes them churn about in currents. Highs and lows move this way and that over the oceans and land masses. Typhoons spring up. Droughts occur. Unseasonal frosts advance. Unpredictability is added to the regular march of the seasons, and life has had to evolve to be reflexive to the predictable and resilient to the unexpected.

Another implication of the tilt is that the globe can be marked off

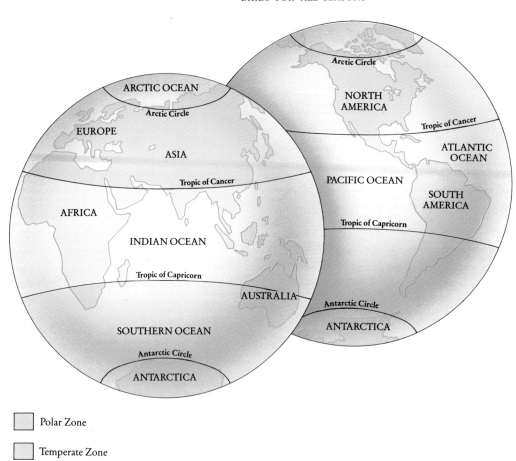

Polar Zone

Temperate Zone

Tropical Zone

into zones: tropical, temperate and polar. The first is the only single zone – the other two exist in pairs – and it is marked at its northern and southern extremes by the circles on the planet where the sun is overhead for at least one day a year. The temperate zones extend north and south from the tropics to the Arctic and Antarctic Circles, where there is at least one day a year of total darkness and one totally without darkness. And the polar zones are what are left. The seasons in all of the zones are influenced by the tilt, but they are different in type and number.

In the tropics, where the tilt itself is noticed the least, the winds which the tilt has created are the most important factor. They shift as the angle shifts, sometimes bringing rain in alternation with dry spells, or never bringing rain or never ceasing to bring rain. In some places in the tropics it is always dry and in some it is always wet, but these are effectively permanent seasons, and the seasons of the tropics are thus these two: wet and dry.

In the temperate zones, the regions near the tropics are much like the tropics and near the poles much like the poles, but in a large belt in between – the cool-temperate zones – there are four distinct seasons: spring, which in the north begins on 21 March, the vernal equinox, when the tilted axis is exactly side-on to the sun, and ends on 21 June, the summer solstice, when the North Pole is tilted toward the sun. Summer ends at the next equinox – the autumnal, 22 September – and winter runs from the time when the North Pole is turned away from the sun, 21 December, to the time when the earth is side-on again. In the southern hemisphere, everything is the opposite.

In the polar regions, the seasons become mixed up with the days and nights, so that at the absolute poles there is only one day and one night, each six months long. The seasons there are light and dark.

To all eight seasons – wet, dry, spring, summer, autumn, winter, light, dark – life has made its adjustments, its cycles, its rhythms. And of all forms of life, few are more influenced than the birds: those higher, warm-blooded animals which can inhabit the land with the mammals, the sea with the fish and the air pretty much on their own.

This book, and the television series of the same title, is intended to convey some feeling of the relationship between birds and seasons, their most important natural control, as they adapt to ever-changing conditions across the globe.

'... the wise thrush; he sings each song twice over,
Lest you should think he never could recapture
The first fine careless rapture!'

Robert Browning

Chapter One
A LITTLE BIT OF BREAD AND NO CHEESE

The earth is now more or less sidelong to the sun and the day and the night are more or less the same length. In the tropics, where the day and the night are *always* more or less the same length, the concern is with the shift in the prevailing winds, and the approaching or departing rains. In the polar regions, the sun has either appeared for the first time in months or has disappeared, but both places are still and cold. In one of the temperate zones — and if this is around 21 March, this will be the southern zone — it is autumn. In the other, and particularly in the cool belt of the northern zone, it is spring and the birds start to sing.

To virtually every form of life that exists here, spring is the first time for many months that nature has shown any signs of letting up, of becoming more benign than hostile, and it is a season to be seized for the urgent, delicate and exposed task of reproduction. Within the first couple of days of the first late-winter thaw, the first small spring flowers have appeared, the first perennials are breaking ground, the first bees and other insects are searching groggily for nesting chambers, and birds that when last seen were fluffed up against the cold and hopping around on the frozen earth as if they were only waiting to die (and many have died), are suddenly back in the trees and bushes, singing for their lives — that is, for a mate and every inch of territory they can get and keep.

For these are the two, and the only two, major reasons that songbirds sing, and in almost every instance the singing songbird is a male. It is his job — before mating, before nesting, before the eggs are laid, incubated and hatched, and before the chicks are fed and fledged — to stake out and defend a territory, a resource base. He usually does this by broadcasting the fact of his existence to other males of the same species, the only other creatures in the neighbourhood likely to want exactly the same resources that he wants.

A male blackbird, for example, will have a good idea of how much space he needs, and will defend it from a perch. At the same time, in the

At the earliest hint of spring, the first priority for most male songbirds is to find a good perch and, from there, to stake out a territory — by singing. The second priority is to attract a mate — also, usually, by singing. The proportion of effort devoted to the one or the other varies from species to species. The simple songs of the blackbird (opposite) and yellowhammer (left) are probably biased towards territory. Those avian Carusos, the skylark (below right) and the nightingale (below left), are mainly lovers.

bush below the tree, a European robin might be establishing his distance from other European robins, but using his kind of song. In either case, if the song is not considered 'good enough' by the other birds of the same species, it will be taken, probably rightly, as a sign of general physical inferiority, and the territory will be invaded.

If, on the other hand, it is a match for the species' highest standards, the singer will not only guarantee himself a safe boundary, he might find his rivals edging rather farther away than they absolutely have to and leaving him more territory than he absolutely needs. This makes him better-fed, more likely to attract a fit mate – for at this point the females are flying from territory to territory on an acoustical shopping trip – and more likely to sire offspring who will themselves be strong songbirds and good singers. And so over time songs will gradually develop and 'improve'.

Songbirds, by which we mean nearly half of all the bird species that belong to the Passeriforme suborder, the Oscines – also called 'perching birds' – are not the only birds that are territorial. Indeed, almost all animals of any kind (and plants, too, in their way) are territorial in some respect. But in an average temperate habitat in the early spring, the male songbird is predominant. His life-cycle is such that he must now make his declarations most emphatically and his evolved way of doing it represents a musical ability unparalleled anywhere else in nature.

For one thing, the songbird possesses a superb instrument. At some point in the long evolutionary voyage from reptilehood, it developed a double voice-box, the syrinx, which instead of sitting at the top of the windpipe, as the larynx does in almost all other vocalising vertebrates, is stationed right at the bottom, where the windpipe divides into the two bronchial tubes that lead to the lungs, and the two openings to these tubes are each covered by an adjustable elastic membrane, the tympanum. As air passes by, in either direction, each tympanum can be made to vibrate and create sound. The bird's brain can control each tympanum independently of the other, tightening or loosening it to alter volume and pitch.

Opposite each tympanum is a 'bump' of erectile tissue. These also operate independently to alter the volume without changing the pitch – much like the mute on a trumpet. What this means is that songbirds can produce two notes and two degrees of loudness at once. They can produce a long note and a short note, one modulated, one not. They can sing in chords and, if they wish, modulate one half of the chord and not modulate the other, something even a pianist can't do. They can even sing two songs at once, though few do. But they do make tunes and trills and note changes that are so subtle and so fast that humans can hear them in their fullness only if they record them and play them back in 'slow motion'.

There are about 4000 true songbirds, each with a different repertoire. The songs can be as basic as the cuckoo's famous springtime two-note ditty, or as virtuosic as the 103-note phrase that a wren can pour out in less than nine seconds, and the degree to which the intentions are territorial or amorous range from purely one to purely the other. Great tits, for example, pair up before the male starts to sing, and so normally his song is only meant for other males. Cock sedge warblers, migratory birds, arrive in northern Europe after a long flight from Africa and immediately establish a territory, not by singing but by chasing other cocks away. Only then do they begin to sing, and this they do non-stop until a mate is installed, at which point they go silent again. Robins, blackbirds and most other songbirds sing for both reasons, and they sing more or less constantly, expending almost as much energy on song as they do on flight. A male yellowhammer is estimated to chant 'a little bit of bread and no cheese' 3000 times a day and half a million times in a season. Look at a singing yellowhammer through binoculars and you can see how he puts everything into it, his whole body quivering with the exertion.

Generally, the more complex a song, the more likely it is to be sung mainly for the purpose of mating. Mating serenades, which logically ought to contain great flights of braggadocio on the subject of genetic fitness, do seem to be harder sung than those which are only saying 'here I am' and 'keep your distance'. If the purely territorial great tit has the bad luck to lose his mate, he responds by singing the only song he knows, the territorial one, but at six times his normal rate.

The most beautiful singers – and the world's ornithologists meet every so often and vote a top twenty – are always more likely to be primarily lovers rather than fighters. Also, the serenaders at the top of the list tend to be the birds that would be hardest for a potential mate to see, which obliges them to paint a picture of themselves musically. The champion British singer – that avian Cyrano, the nightingale – is nocturnal, and the *world* number-one has to contend with the perceptual

Many birds, including these South-east Asian pheasants, use their feathers to the same purpose that songbirds use song: threatening males and attracting females. To this end, the feathers of the Himalayan monal (opposite) refract sunlight to produce a metallic effect. Temminck's tragopan (left) lowers a blue and red 'bib'. The silver pheasant (below left) produces a spectacular feather display entirely in monochrome; and the golden pheasant (below right) uses nearly every colour there is.

handicap of living in dense, dark forests. He does this not only with a spectacular song — one that actually overcomes the blanketing effect of the vegetation — but also an extremely 'loud' visual display. This is the superb lyrebird of Australia.

Purity and complexity — two qualities which make up a large part of the human judgement of beauty in birdsong — are, along with repetitiveness, important strategies in the development of musical muscle. Another, rather more obvious one perhaps, is volume, or carrying power. Male lyrebirds can be as near as five metres apart and, unless they are displaying, still not be able to see each other, but because of the sheer volume of their arias, they are seldom nearer than a thousand metres. Other birds in more open habitats — and almost any habitat would be more open than a lyrebird's — can be more subtle, enhancing their carrying power by, for instance, going to the highest possible perch. A mistle thrush will always sing from the top of the tallest tree, and even in landscapes without trees, birds such as skylarks have evolved to deliver their serenades from mid-air. Meadow pipits and rock pipits go a stage further, performing a choreographed flight as they stake their claims and present their sexual credentials.

Pipits dance as they sing, and the beautiful lyrebird is also a beautiful singer — for acoustical display and visual display are essentially the same. At nesting time, the purpose of almost any male bird anywhere is to outdo rivals and to gain a mate, and if prevailing conditions in the bird's niche favour sight over sound, then the bird will rely more on its feathers than its voice. In fact, the other half of the class Aves, the birds that aren't songbirds and have larynges instead of syringes, are mostly visual displayers. Many of the larger birds, such as the galliformes (otherwise known as game birds, a name which makes human intentions towards them rather transparent), put together elaborate shows that combine sound, dance, colour and pattern.

Among the most brilliant and persistent of these are Temminck's tragopan, a native of China, and the satyr tragopan of India and Nepal. In each species, the sexually excited male lowers a bright red, white and blue bib, as if he were letting down a small window-blind, and then raises two matching sky-blue horns, before launching into an elaborate clicking, prancing demonstration for the speculative hen. Another pheasant, the Himalayan monal, displays in a dazzling flight over the great mountain range's high cliffs and deep, forested valleys. And the

Chinese golden pheasant not only lays out its splendid golden train and lowers a fabulous gold and black collar over its scarlet breast plumage (Europeans once thought that birds like these were figments of the imagination of Chinese painters), it will also engage a rival in real foot-to-foot combat. This is an unusually direct approach for a bird that has taken such evolutionary pains with its display, for display is widely regarded as a way of establishing genetic fitness without anyone having to get hurt.

Farther north, where the vegetation is less dense, the pheasants, quail and other fowl have had to sacrifice sheer splendour for an element of camouflage, but in this case subtlety is relative, and of all the birds in these regions, the galliformes still have the most striking displays.

The American wild turkey, puffed up to nearly twice his normal size and with his bright red wattle and his elegant mahogany and amber tail-fan, so caught the imagination of the new arrivals from Europe that he was seriously considered as the national emblem of the United States (but was narrowly beaten for the honour by another bird, although the turkey remains emblematic of Thanksgiving, one of the two most important national holidays). Certainly, an informal symbol of the grasslands which once formed a broad, uninterrupted swathe down the middle of the continent, is the greater prairie chicken, or pinnated grouse. In display he looks as though he has wrapped himself in an ermine stole, and from head-on, with his tail-feathers rising behind – so rigid that they look as though you could make them twang – he seems to be wearing a huge star-shaped crown.

Though the basic need for territory and mate is almost universal among birds, the methods of securing them are so various that almost any strategy you can imagine is practised by some species somewhere. Among the many factors that determine exactly what a bird does at this point in its cycle is whether, like most land birds, it nests apart from other members of the same species, or, like most seabirds – which have vast ranges and large populations but very little in the way of a landhold – it nests in colonies. Seabirds have their own versions of all the things that other birds do, but in defence of a patch scarcely big enough to turn around on. As 'city-dwellers', their concern is for a mate – which they generally take for life – and a level spot for an egg or two. For food resources, which are normally plentiful, they commute.

It sometimes happens in bird species that the concern is about 99 per

Many birds that have lifelong pair-bonds begin their 'marriages' with an elaborate courtship, complete with dances. A male gannet (top) raises his wings and beak in a 'sky-pointing' display. Wandering albatrosses (centre) show off their three-metre wingspans. Red-crowned cranes (bottom) have a love-dance combining pattern and shape with musical movements. The courtship of laysan albatrosses (opposite) includes 'necking'.

cent mating and 1 per cent territory. At such an extreme ratio, the territory becomes abstracted into a useless bit of ground known as a 'court' and is defended during the ritualised 'lek', when all the males of a species get together and hold something like a puffing, dancing and calling contest. In this ceremony – perhaps most familiar to Europeans as practised by the ruff, and to Americans by the prairie chicken – the males stand on their pretend territories and work each other up into a frenzy, which then culminates with the females choosing their favourite performers. These are usually brilliantly conspicuous affairs, with the birds abandoning all other considerations (and they must occasionally have a fox or a coyote rubbing its eyes in disbelief).

However, there is also such a thing as a nocturnal lek, based mainly on sound. One of the most recently discovered of these is the remarkable communal display of the New Zealand kakapo. Not satisfied with being one of the few temperate parrots, one of the two ground-nesting parrots, the only flightless parrot, the only nocturnal parrot and the largest parrot (as well as the most endangered parrot: it has been written off as extinct more than once, and there are still less than twenty known individuals), the kakapo is the only parrot that has a lek, and it is a lek so unusual that some people wonder if another name ought not to be invented for it. The males select widely spaced spots along a ridge over, preferably, a valley with lots of sound-carrying potential. They then dig shallow 'bowls' in the dirt, settle in and deliver steady, resoundingly loud 'booms' into the night, stopping occasionally to shuffle about in their bowls in an awkward little wing-flapping dance (just about the only thing kakapos use their wings for). The females, in the meantime, parade back and forth along the ridge as if on a late-night shopping expedition.

Another nocturnal lek is the one held by the great snipe of northern Europe and Asia, and it, too, is based on loud calling. Otherwise, leks happen in the daytime, and the species that indulge in them include great bustards, some of the hummingbirds – which perch side by side on tiny branches and hold marathon sing-ins – and several tropical birds, among them cocks-of-the-rock and certain birds-of-paradise.

The general rule about birds which display, court and mate in this way is that, because the females have taken over the entire responsibility for attending the nest and because the chicks hatch with a precocious ability to feed themselves, pair-bonding is unnecessary and territoriality means very little, at least to the males, which once they have contributed their genes are of no further use to anybody but themselves. That means that for a male the be-all and end-all of reproduction is to stand his ground at the lek every year and inseminate as many females as will have him, while the females' interest is in being inseminated only by the very best males. And the conflict of these two interests has led to the evolution of some birds that are very, very spectacular, either visually or acoustically or both. It is also worth noting that the lek display is not a primitive characteristic: monogamy came first.

And monogamy probably came before yet another system, one that is even less common than the sort of polygyny – one male serving several females – characterised by lekking. Some species practise polyandry, and then it is the female which is either footloose or attended by a retinue. For example, a female Galapagos hawk or a female Harris's hawk of the North American deserts can have as many as four mates, all bringing food and helping her look after the eggs and young. In the Andalusian hemipode and the painted snipe, the larger and more colourful female displays, courts and fights, and when she eventually lays her eggs, she leaves her mate sitting on them while she goes off in search of another accommodating male.

Most of the ratites – the big running birds – exhibit some variation of polyandry, often at the same time as exhibiting variations of other mating systems. Cassowaries form pairs in much the same way as monogamous birds, but once the eggs are laid, the female, like the female hemipode and snipe, leaves the male on the nest and goes to another male. Emus do the same, but in this case the females simply leave, and do not mate again. The female rheas of South America mate, lay, leave and look for a new mate, but with this interesting twist: they do it in groups, as intact harems. Like many other polygamous birds, the male puts on a display and attracts a harem. The members of the harem all lay in the nest, and then, all together, they get up and move on to their next sultan.

But it is the biggest of the ratites, the ostrich, which has managed to mix all three concepts. The males display in the same general, if necessarily grander, manner as many other birds. Each one stakes out a territory containing a scrape ready for egg-laying. He announces the fact by showing off his neck – which has turned either bright red or bright blue, depending on species – and by booming regularly and standing as

Male ruffs (opposite and below) have more variation in the colour of their plumage than any other bird, all part of the requirements of their 'lek' courtship. This involves standing in a gathering of other males and mounting a spectacular display for the benefit of attending females, who will mate with the male that is dominant. The males of some species, such as ring-necked pheasants (left), can actually come to blows.

high as he can, virtually on tiptoes. Trespassing males are evicted by fearsome, hard-running charges but, when a female enters his territory, he trots towards her, holding his tail high. Then, suddenly, as he is nearly in front of her, he drops to a squat and starts to sway from side to side, snaking his neck around and waving first one wing in front of her and then the other. He can repeat these motions scores of times while she looks him over. When she is eventually convinced, she assumes the submissive posture by lowering her head and wings, and he raises his wings above his back and mounts her.

If this is his first female of the season, she becomes his 'major' mate, lays the first egg in the scrape he has prepared, and continues to lay there about once every two days. But the male, meanwhile, carries on displaying, attracting and copulating with more females. When these 'minor' hens are ready to lay, they are led to the scrape, where they deposit their eggs. Then they leave to become the minor hens of another male – and only the major hen stays to do the incubation.

Few mating systems could be more complicated than this, or more frenzied than the average orgiastic lek, but for most birds the essential business of establishing a territory, attracting a female, courting and copulating are rather more sedately paced. Our blackbird, for instance, will attract females with the same song that deters males, and over a period of days may thus entertain several hens who have stopped in to look him over and to listen to his song before moving on to another male's territory. This more or less applies to most members of the thrush family, which also includes both European and American robins, night-ingales, fieldfares, forktails, bluebirds and the great range of birds that actually go by the name of thrush. Finally, one female will find the other males either inferior or already mated, and will return more and more often to our male's patch, until at last she stops leaving. At this point the pair will copulate for the first time, and immediately begin to build a nest.

Blackbirds and the other thrushes are rather sober citizens when it comes to courtship, and probably the only special behaviour they exhibit is limited to chasing, mutual feeding and mutual preening. Other birds, particularly larger and longer-lived ones and the ones that mate for life, perform wonderful courtship dances in the air or on the water. One of the most sensational of all is the famous water-dance of the western grebe. In perfect syncopation down to the angle of the curve in their necks, the attitude of their beaks to the sky, and the way they hold their wings out behind, the couple will run back and forth across the surface of a lake and then suddenly dive and continue the ballet under water.

Most birds of prey will spend a day or two in acrobatic swoops and soars, sometimes clasping talons and folding their wings so that they hurtle earthwards in a double sky-dive. Clattering their bills, storks stand breast to breast and throw their heads back as far as they will go, turning the world upside down. Herons and albatrosses point their beaks straight up, spread out their wings and turn in a slow circular dance. Adolescent cranes of every species hold something like a spring ball, when after a few days of elaborate prancing, beak-clicking dances, they eventually pair off with their lifelong partners.

For most species the courtship ritual is also when the first copulation takes place. It is a very brief act, with the male climbing on the female's back, touching his cloaca to hers and then, in a matter of seconds, climbing off. Some birds actually stay celibate for a while after they have formed their pair-bond, as if they were having a long engagement. The longest such period is probably the year that albatrosses let pass between their bonding and its consummation.

Among birds, monogamy is the rule. About nine-tenths of all species practise it, for periods ranging from a single season to life. This is primarily because most of the more difficult reproductive tasks can be better performed by two than one – and increasingly better performed by two with the experience of working together: building complex nests and feeding demanding chicks that can neither fly nor forage for themselves. There is also a certain amount of time and energy saved by not having to search constantly for a mate.

Birds, like all organisms, operate very close to the margin. Evolution is a perpetual thrust towards getting the most out of nature for the least expenditure of energy, and for most of the tasks performed by a warm-blooded animal which uses the air as its main transport medium, the nest as its home and the external egg as its reproductive unit, two can do together what neither could do alone. Each saves the other just enough energy to bring them both to the possible side of the margin of existence.

A sage grouse (top) entices a female with his crown-like tail. The frigatebird (above right) uses a vivid frontal display. In a unique photograph. a New Zealand kakapo (above) shows off his wings and does a shuffling dance in an area where all the female kakapos have died out. A superb lyrebird (right) is lost to view beneath the equivalent of three tails: wiry. filigreed and lyre-shaped. The peacock (opposite) fans out what may be the most elaborate visual display in the animal kingdom.

'The nightingale, the organ of delight,
The nimble lark, the blackbird, and the thrush,
And all the pretty choristers of flight,
That chant their music notes in every bush,
Let them no more contend who shall excel;
The cuckoo is the bird that bears the bell.'

Anon

Chapter Two
THE WHOLE CUCKOO FARM

The short stretch between the end of courtship and the beginning of laying – about three days – is just about the time it takes for our two blackbirds, working flat-out during the daylight hours, to put together their rather large nest. She builds and he forages for both of them, and their nest, like few others, always looks as if it has been thrown together in great haste. That is, until you see the business part of the nest – the bowl – which is smooth, dextrously woven, packed with mud and lined with grass stems.

The inside of the nest looks comfortable, and it is a matter of some amazement that it was put together with the random materials whose other ends are poking, dangling or drooping over the edge of the branch or eave joist that is holding the whole structure up. It could even be that the sloppy appearance is part of the design, a camouflage for a nest that is usually built in the relative open, compared with the neatly constructed robin's or wren's nest hidden deep in the shrubbery. All of these nests are at about the mid-level of structural sophistication but could never have been built in time by one bird working alone. 'Higher' nest-builders give domes to their nests, a few going so far as to hang them from the ends of branches, such as the African sociable weaver. 'Lower' ones are still earthbound.

Some birds, lek species for example, very rarely have anything more than a scrape on the ground or a clearing in the grass, and the same holds true for other birds, that, even if they do not have a lek, still leave nesting and rearing entirely to the female. The fact that *either* parent should attend the nest is still something of an evolutionary step up from the reptiles, which, with a few exceptions, lay their eggs in a hole or under some vegetation and then go about their business, leaving the eventual hatchlings to fend for themselves. No reptiles incubate their eggs – nor could they, being cold-blooded – but some, such as the crocodiles, do stay around to protect the eggs from predators, and this

Every species of bird has devised some way of keeping its nest safe from predators. The feathers of the fiery-necked nightjar (right) are barely distinguishable from the surrounding leaf-litter. Some hummingbirds, in this case the ruby-topaz (below), 'glue' their nests to branches that are too thin to support any possible non-flying predator. The cliff-ledge nests of kittiwakes (opposite) are impregnable to anything but winged enemies.

kind of behaviour might have represented the first glimmerings of the attended nest. Birds, though, are doing more than just baby-sitting. They have developed internal heating, which like all technical advances has a catch: the embryo must be kept at a steady temperature. The need to build nests and incubate eggs is perhaps the step backwards for the two steps forward conferred by warm-bloodedness, which allows birds to inhabit parts of the planet that are generally too cold for reptiles.

The first incubating birds must have behaved a bit like the stone curlews which, with almost no preliminaries, lay their eggs directly on the ground and then sit on them. The evolutionary advance here is the camouflaged eggshell, which allows the mother to get off her nest occasionally to feed. Ringed plovers, many terns, sandgrouse, night-jars and bustards all 'nest' this way, and little terns, which lay their eggs on beaches, are nearly as basic except that they build a low wall of sand around the clutch. The next step in nesting sophistication must have been associated with the development of flight and the birds' realisation of its advantages: the use of elevated sites such as cliff ledges, crevices, caves, walls and trees. Some of these can be as basic as the unadorned ledge nest – the kittiwake's, for example – which, however simple, does keep the eggs away from gravity-bound predators. There are even a few species of ducks – normally ground-nesters, because their chicks are expected to leave the nest as soon as they are hatched – that have taken to laying in the hollows of trees, which of course means that the day-old ducklings have to begin life with a ten- or twelve-metre plummet.

Most reptiles dig holes for their eggs, and so the instinct to alter the immediate landscape was already there when the first birds evolved, and they soon graduated to the excavation of scrapes and burrows, using their beaks as picks and their feet as shovels. Sand martins and bee-eaters burrow into sand cliffs, kingfishers dig into earthy riverside banks, and an unusual bird called the crab plover nests in an unadorned depression at the end of a tunnel that can be two and a half metres long. The gila woodpecker of the Sonora desert in the south-western United States drills into the tall, juicy saguaro cactus, which gives it a home that is not only safe but water-cooled. It has no further embellishments, though, and the general rule about all of these burrows is that wind, rain and temperature extremes are kept out, and no more construction is needed.

One, more specific, explanation for the origins of nest-building is that they grew out of a bird's habit of venting frustration – at, say, losing a territorial dispute – by what behaviourists call 'sideways throwing', picking up bits of vegetation, pebbles or whatever else comes to beak and flicking them to one side or behind. If these tokens were to accumulate around a particularly frustrated bird's laying site, their protective value – as a windbreak or a marker, if nothing else – would begin to be appreciated, and over the aeons the purpose of the throwing would change. Indeed, certain nests, including those of redshanks, swans and many gulls, are still cobbled together with no more sophistication than 'sideways building'.

Lapwings and other waders improve slightly on the windbreak concept by also making a small platform to keep the eggs dry. Flamingoes build tall mud cones to keep their eggs out of the water, and water rails construct their platforms out in the marshes, not because there is no dry land available, but because a nest in a marsh is out of reach of most mammalian predators, mainly foxes.

The nests of grebes, which also use water as a defence, make the final break with *terra firma* and are actually rafts of vegetation moored to the reeds. And grebes, in designing these egg-boats, appear to have stumbled upon one of the more unusual functions of a nest: as an incubator. These birds, which lay white eggs, cover them with a layer of vegetable matter. This of course gets wet, and when it does it conducts, and holds, the sun's heat. Meanwhile, the material on the underside of the nest is rotting, and that makes heat, too. It is not so consistently warm that grebes can avoid doing their fair share of personal incubation, but when they leave their nests they can spend more time feeding, knowing that their central heating is on at home.

Perhaps the best studied of these nests is the black-necked grebe's, which looks like a thickish but flat disc sitting on the surface of the water but which on examination turns out to have a ball-shaped hull that is deeper than the nest is wide. This kind of structure has gone some distance from those of birds that toss a bit of detritus in the general direction of their eggs. But the next step, actually taking the nest up off the ground or the surface of the water and into the bushes or trees or cliffs or mountain peaks as a fully constructed home, puts the endeavour into an entirely different dimension.

Again, however, progress has meant complications. Simple ground predators are neutralised, but trickier flying and climbing ones, such as

Among the amenities nests offer their owners is temperature control. The little island built by the Slavonian grebe (opposite) gains some warmth from the rotting vegetation used in its construction. The gila woodpecker's desert abode, inside a saguaro cactus (left), is water-cooled.

snakes or hawks, have to be accounted for, as do other perils of high-rise living – for example, the physical reaction of a tree to a hard wind. As threats decrease in number they seem to grow in sophistication, and birds can demonstrate a comparable ingenuity in their use of physical principles and natural materials. The wind-in-the-tree factor can be accommodated by building, with masterly skill, firmly secured deep-bowled nests, domed nests or penduline nests that hang as straight as plumb-lines regardless of the tree's angle to the earth. These latter reach their extreme in the elaborate, multi-doored nests of Africa's sociable weaver. For potential predators to have any success they first must find hidden entrances, penetrate thick bushes, brave the presence of thorns or more animate dangers (fierce ants are a favourite) or, indeed, locate the nest at all.

Some birds even use chemical defences. North America's white-breasted nuthatch builds its nest of leaves inside the hollow of a tree, and using adobe – reinforced mud – plasters up the entrance until only a nuthatch-sized opening is left. Then it goes out and finds a noxious blister-beetle, spears it and paints its juices around the edges of the hole. This stinging liquid deters everything except, of course, white-breasted nuthatches, which are immune.

In general, though, defences are structural and are erected against the larger predators, such as foxes, snakes and birds of prey. The organic threats at the other size extreme – the lice, mites, fungi and even smaller organisms that invade all birds' nests and seemingly make all birds' lives a misery – are controlled or avoided by other means, usually basic cleanliness or a predisposition to build a new nest after each brood leaves home. But a few birds use chemistry against the smaller predators, too. One of these brings home the green, aromatic leaves of the margosa tree, which produces chemicals that somehow prevent insects and other arthropods from laying eggs. It is a property also known to the local people, who sew the leaves into their clothes, as permanently installed mothballs. As it happens, these people live in northern India, the only place where the tree grows, and the bird in question is *Passer domesticus*, the house sparrow. It has only been observed doing this in India, but who can say that in other parts of the world it isn't also choosing, if less conspicuously, nest materials according to their fumigatory properties? Certainly another common bird, the starling, is known to adorn its nest with fresh leaves and blossoms of plants that inhibit the development of lice and bacteria, and many birds of prey, which, like starlings and sparrows, tend to settle permanently in one spot, also continually renew the supply of various fresh sprigs around their nests' rims in their perpetual repairs.

And there may be nest-building considerations that are neither structural nor defensive. It is not clear whether birds which adorn their nests with eye-catching objects – most typically, the members of the crow family – are exercising some kind of avian aesthetics. Nor is it clear what *other* reason they would have. Magpies are so famous for the practice of 'collecting' apparently useless human artefacts – anything slightly shiny, from bottle caps and bits of foil to metallic legal tender – that the bird's very name has come to describe a human with a similar tendency. One theory, with regard to the feathered magpies at least, is that they are indulging in some kind of sexual display and that this is the rather primitive end of a practice that reaches its logical extreme with the bowerbirds of the Antipodes.

The structures erected and defended by the males of the several species in this family have become so lavish and intricate that they are often taken for nests. They are displays, like the lyrebird's tail, and once a female has mated with the architect of her choice, she goes off and builds a nest of her own, a simple cup in a tree. The males, meanwhile, keep on weaving, refining, collecting, decorating, smoothing, stacking, even painting (they will dampen a wad of bark with their saliva, dip it in charcoal dust and touch up the doorways), stealing from each other and, in the case of the satin bowerbird, persevering in the eternal quest for anything blue – in particular, anything blue that is also a feather. Somehow, this has become prime currency, and a male satin bowerbird's ability to find blue feathers (not of the same shade as his own) is the best indication of his genetic fitness. Ornithologists who have compiled statistics on male satin bowerbirds' success rates (and life is tough in a bowerbird population, where only about a quarter of males ever build well enough to persuade any females to accept their genes) have found a direct relationship between the number of matings and the number of blue feathers.

There is no doubt that the thrush-sized, often drably coloured bowerbirds build as a kind of sexual display, and the finished products are, without question, the Taj Mahals of the natural world. Though the nests vary greatly from species to species, they are usually creations of

Some birds are capable of prodigious and extremely delicate feats of construction. The great hotels built by African sociable weavers (near and far right) feature downward-facing doorways to confuse predators, usually snakes. The male satin bowerbird (below right) builds an intricate arena trimmed in blue, not for nesting, but as display. Grey-headed albatross chicks (opposite) are kept off the muddy ground by elevated nest-cones.

great delicacy and cleanness of line. They can be very simple, like the tooth-billed bowerbird's arena, which is nothing more than a small circle kept clear for an arrangement of the bird's collection of colours and shapes. Or they can be highly labour-intensive, like the twin 'maypoles' of the tiny golden bowerbird – a pair of columnar structures that can reach three metres or more, built of thousands of twigs around a pair of conveniently proximate saplings and connected by a sort of threshold carpeted with moss and lichens. The regent bowerbird clears a pair of 'avenues' leading to either end of a frondlike tunnel, persuading prospective mates to enter by carefully arranging his hard-collected objects on either side of the pathways.

The human reaction to these edifices is not only one of sheer amazement that 'lower' animals could construct something so arcane and complex, but of wonder and appreciation at their beauty. And indeed, the shiny, man-made objects that catch the eye of the magpie are often made the way they are to catch the eye of a person. And so the question of whether the magpies are living up to their name because of aesthetics or because of sexual display begs a further question: what's the difference? Cannot a bird be applying aesthetics in order to attract a mate? Or, for an object or an act to qualify as aesthetic, must it be free of an ulterior motive? If so, what about people who create art in exchange for money? For that matter, what about people who write love poems in order to attract a mate?

The human race does not claim uniqueness in the need for food, temperature control and a bit of territory – and the pleasurableness of sex. The old-time 'abilities' that are supposed to distinguish us from the rest of creation – tool-making, language, a capacity for happiness and remorse – are continually being discovered these days in other animals, and so why should we continue to arrogate to ourselves an ability to appreciate beauty? It is quite possible that a magpie picks up a metal button and takes it to its nest because the button is, as its manufacturer intended it to be, pretty. And if it is a male magpie, perhaps he will hope that his mate will also find it pretty and not leave him for another magpie (the 'divorce' rate among certain monogamous birds has been estimated as about one for every three pairings). If a female bowerbird thinks to herself: that is the most *beautiful* construction of sticks and moss and berries and blue, blue feathers that I have seen all week, and he's taken so much trouble with it all . . . does it make her assessment any less valid if it also serves a genetic function? Does our own aesthetic sense have a genetic function? Certainly, to look at our phenomenal reproductive success, you would have to think that *some* genetic factor is working well – too well.

Evolution is very accurately represented by our most usual analogy for it – the tree. The tips of the branches may be as high as each other, and it may have taken each as long to grow, but they are in different places. 'Progress' in evolution – the gradual change that allows an organism to live in a place and in a way that no other organism can live – is usually achieved by travelling farther than anything else along a particular line, by becoming ever more specialised. The best nest-builders, for example, would not necessarily excel in any other avian skill. Some birds can fly longer distances than any others. Some can fly faster. Others can swim and dive well. Some have better eyesight. Some survive largely on their talents at deception or camouflage. Some survive by virtue of having evolved in places where it is easier to survive, most notably islands without any natural predators (a piece of luck for which island birds pay dearly when predators – usually man and his dogs, cats, rats, goats and pigs – appear, as it were, out of nowhere, for most of the birds that have become extinct in the past few centuries or are in imminent danger of becoming extinct now are islanders, usually flightless, usually suicidally friendly: in short, dodos).

So, a basic nest-builder such as the nightjar, which quite simply builds no nest at all and sits on its eggs wherever it has decided to lay them, is in no sense inferior to, say, the magpie with its sturdy, domed, dry, unassailable nest high in the treetops. Instead, the nightjar has concentrated on coming as close as an animal can to invisibility. Its camouflage conceals the bird and its eggs on a certain kind of heathland, and it saves its most active feeding for dusk and dawn when, it could be argued, one set of potential predators has gone to bed and the other hasn't quite woken up. More importantly, it feeds on insects that are busiest at the same times of day.

As nice as it would be to invent one, there is no single line of nesting strategies that starts on the rocky shore and ends at the top of the tree. At every point of development – and remember that there are nearly 10,000 species of birds, each with its peculiar way of making a living – is a junction from which several branches emanate, leading birds to become master builders or masters of disguise or deceit, or to be bad at

One of the most common nesting strategies, and one of the safest, requires neither building nor camouflage; just finding a hole in a tree and moving in — as this screech owl (opposite) and South American chestnut-mandibled toucan (left) have done.

all those things and particularly good at something else, perhaps even to become the predator that all the others are trying to hide from or outwit.

The house sparrow is an example of a bird that became very successful by mastering the art of picking the right friends. Certainly in the past nine or ten millennia, no animal could have done better than to hitch a ride with the burgeoning *Homo sapiens*. Rats and cockroaches have done rather well too, but sparrows have managed it usually without making pests of themselves. They share our roofs without getting into our rooms, and their protective camouflage does not make them invisible so much as eminently unremarkable. Of course there were house sparrows before there were houses; before, even, there were humans. But when humans and houses came along, house sparrows were ready with talents refined on other species: one of their prehistoric nesting sites might have been the lower storeys of the great, messy nests of eagles.

Wood pigeons, relatives of another of our bird hitchhikers, have also been known to use eagles' nests. Black-necked grebes are almost always found living under the protection of black-headed gulls, which extract a sort of tax from their tenants, not by preying on them personally, but by stealing a certain proportion of the food they bring home. That the grebes find this arrangement convenient is borne out by the fact that when the gulls, as gulls do, pick up sticks and move to another nesting site, the grebes wearily follow.

Other birds can have their food-searching abilities exploited by black-headed gulls. Lapwings and golden plovers are two of the most common, but the gulls don't seem to give them anything in return. For the grebes, the gulls drive away skuas. They also do this for eider ducks, and the ducks have become so dependent on *somebody* doing it that one long-term colony in the Arctic has set up alongside a research station's huskies. And there is at least one instance of people trying to use birds for protection. In parts of New Jersey, farmers are in the habit of putting up poles topped by old wagon wheels for ospreys to nest on. In turn, the ospreys are said to keep other birds of prey away from the farmers' chickens.

All these are examples of one degree or another of symbiosis. Depending on how you look at it, this is mutual aid or mutual parasitism, but it is nevertheless an expression of the vital necessity, throughout nature, to save energy. One of evolution's hard truths is that the organisms that have survived are the ones that in their given circumstances have got the most for the least, and were the most energy-economic. Symbiosis is saving energy by getting another organism to do for you what it does better, while you do for it what you do better. The balance, however, is not always equal. Sparrows, while not bothering us, still take more than they give. In fact, they give virtually nothing and so represent the very beginnings of parasitism proper; the respectable end of a range of one-sided nesting relationships whose other end is represented by that scourge of bird-dom – the European cuckoo.

What might be called 'cuckooism' – laying eggs in another bird's nest – is by any reckoning the most energy-efficient nesting system of all, and it is practised in varying degrees by several birds besides the ones that actually bear the name. In the bay-winged cowbird of South America it is seen at its most basic – using a nest that another bird has built and abandoned. Bay-winged cowbirds don't build nests, but otherwise they are ordinary parents, incubating their own eggs and feeding their own chicks. The next stage, and the first in which parenting – that activity which probably consumes more avian energy than any other – is handed over to another species, is when a bird simply lays an egg in another's nest and has done with it. An example of this is the black-headed duck, which deposits one egg per nest for other waterbirds to incubate. When the duckling hatches, it can take care of itself, so it simply pops out of the nest and heads for the nearest water.

Other 'cuckoos' which parasitise birds of roughly the same size and shape allow the host to rear and fledge their chicks as well. The Indian koel uses crows as foster parents, and the chick has evolved to look exactly like a crow chick. Only after it has fledged and only if it is a female does the black plumage give way to brown. In Africa, whydah finches lay their eggs in the nests of estrildine finches and, despite stringent evolutionary efforts on the part of the latter to produce marks around their chicks' mouths that will distinguish them from whydah chicks, the whydahs have kept up. The chicks of both species now have an extensive and useless array of coloured marks and bumps, and estrildines are still raising whydahs.

In these instances, though, the host is losing very little. It is simply sharing out a bit of body heat and generally cutting the cake into, say, seven slices instead of six. It is at cuckooism's next stage that the loss suddenly becomes major: host parents have their offspring replaced by the parasite's. The chick of the African honey-guide, which usually

The life-style of the European cuckoo (opposite) revolves around spending the winter in Africa and, in the rest of the year, persuading other birds to raise the young cuckoos. In most instances, cuckoo eggs are a fair match for those of their small hosts, but in the case of dunnocks (top) colour and size don't seem to make much difference. When a cuckoo chick hatches, as this one (centre) has just done in a reed warbler's nest, it first ejects its own eggshell and then gets busy (bottom) on the unhatched eggs of its hosts. In time, this leads to one of nature's saddest spectacles (page 53).

imposes itself on woodpeckers and barbets, is born with sharp hooks on its beak-tips, and it uses these, before they fall off, to kill the other nestlings. But the literally over-the-top ultimate in the art of avian boorishness is the European cuckoo. It first removes one egg from the nest of the host, which can be any of a number of smallish songbirds — robins, meadow pipits, reed buntings, dunnocks, reed warblers — and in its place lays its own. It then thriftily eats the egg that is now spare and flies back to the perch from which it defends its territory of host nests from other female cuckoos. The whole strategy is an odd and highly specialised one, and up to this point — before hatching, before incubation, before the host birds have even come back from their feeding foray or whatever else made them leave their nest untended for a minute — the cuckoo has had to use several odd and highly specialised adaptations.

First, unlike those cuckoos whose chicks simply share their hosts' nests with the hosts' own chicks, the European cuckoo is much larger than its hosts. It is as big as a sparrowhawk, but it still has to lay a convincing-looking small bird's egg — and the result is that in all of nature no other bird is so large in relation to the egg it lays. It has not quite been able to refine the egg-size all the way down to that of a robin or a warbler, but it is very close, close enough to work.

The European cuckoo also has to have an egg that has the same markings as the hosts' egg. This sounds straightforward enough until you remember that there are several host species, all with differently marked eggs. Exactly how cuckoos cope with this problem is, biologists admit, still a puzzle. It appears that a female always lays in the nest of the species which brought her up, and so she would have inherited from her mother the propensity to lay a certain kind of egg: pipit cuckoos always lay pipit eggs, robin cuckoos robin eggs. But does that make pipit cuckoos and robin cuckoos different species? Not at all, because the usual working definition of a species is a grouping outside of which it is impossible to reproduce, and any male cuckoo can and will breed with any female, whatever her host species. Egg colour is carried only in the female line, and not affected by male genes, so a male born in a robin's nest can mate with a female from a pipit's nest without running the risk of having daughters that might lay robin eggs in pipits' nests.

Another cuckoo quirk is that the female is the main territory-holder, standing guard over all her little nest-builders and chasing away other female cuckoos (occasionally one is allowed to stay around if she makes it clear that she parasitises a different species). Since the female lays about a dozen eggs in a season, her patch has to have at least as many host nests, plus a few extra as seed corn. The male, the one that does the cuckooing, shares the territory, singing for both of the usual reasons. But cuckoos neither build nests nor raise their own young, and would seem to have no reason at all for forming a pair-bond. Nonetheless, they are monogamous. On the face of it, it would seem that the male, at least, would be relatively free of the territorial imperative, since the only bird he really needs to look after is himself. He could therefore roam around trying to inseminate as many females as possible, a strategy that usually works to a male's genetic advantage, and one which is employed widely by many species of insects, reptiles and mammals. Instead, the male cuckoo mates for life. Each year, he and his mate return from southern Africa to exactly the same patch they inhabited last year. They quickly court and copulate, and then the female assumes her old vantage point for watching the progress of the nest-building by the robins, pipits or warblers, and the male goes to a perch of his own and sings his boundaries both to other males and to letter-writers to *The Times*.

So, why? Let it be said quickly that no one knows, but there are some considerations. The whole cuckoo family, which can be identified by their two toes in the front and two in the rear (most birds have three toes in the front and one in the rear), comprises 127 species, and only 45 of these are parasitic. The others, including most of the New World cuckoos, such as the yellow-billed cuckoo and the roadrunner, all nest conventionally and monogamously. The parasitic behaviour almost certainly evolved after the normal behaviour, and if monogamy were no outright disadvantage it might have persisted simply as an old, diehard habit. Evolution selects *against* disadvantages rather than *for* advantages, and a lot of creatures retain certain characteristics, both physical and behavioural, for no other reason than that they are harmless.

Another way of looking at the cuckoo's mysterious monogamy might be to broaden the definition of nesting. Cuckoos don't actually gather up sticks and build nests, but they oversee a very valuable population of birds that do. And from the cuckoos' point of view, the little birds are doing it *for* them. They are the cuckoos' possessions. To the cuckoos, the whole territory is a nest, and the pair of them administer it like a farmer and his wife. By employing a set of tricks not unlike the tricks that humans use to get their animals to serve them, the cuckoos have domesticated the other birds, and the males are needed to maintain the well-being and integrity of the territory, the cuckoo farm.

Anyway, the male is not all *that* monogamous. Females that farm different species in overlapping territories will sometimes share the resident male, and unmated females without territories often gather together in the territorial fringes, where there are no good host birds, and enjoy the attentions of the mates of the females that have chased them there, a kind of bird brothel that frequently leads to cuckoo eggs turning up in the nests of blackbirds or goldcrests or house sparrows (European cuckoo eggs have been found in the nests of more than 125 species – including 50 in Britain – and yet 77 per cent of British cuckoos use meadow pipits, dunnocks or reed warblers).

On the odd occasion a pair of these 'unnatural' hosts will actually incubate the egg (instead of wondering what's come over the neighbouring pipit, and chucking it out), which can lead to the pitiful spectacle of an exhausted goldcrest standing on the head of a gargantuan chick and dropping yet more food into its bottomless craw. The end result is a cuckoo that lays pipit eggs but is imprinted on goldcrests, a state of affairs that does not often persist into a second generation. This is not to say that in the future an errant cuckoo might not stumble across a host bird which seems largely blind to the differences between eggs and will sit on anything it finds in its nest and raise anything that hatches. Then, presumably, succeeding generations of cuckoos will use this species, with their eggs gradually evolving to look like the host eggs, on the principle that *sometimes* an egg will be noticed and evicted and that the cuckoos with the best chances of survival are the ones that hatched from, and will lay, the eggs most closely resembling the host eggs.

It is also possible that a cuckoo will discover a species that really cannot tell the difference, under any circumstances, between its own eggs and those of another cuckoo host species, and then there will be no incentive for the eggs to evolve. This appears to be the case with the species most commonly parasitised by British cuckoos: the dunnock, or hedge sparrow. Its own egg is a clear blue, but the egg that the cuckoo leaves in its nest is cream-coloured, with grey mottling, not very different from the one it leaves in reed warbler nests. Dunnocks, it appears, just do not discriminate. Unless – an afterthought – cuckoos have only recently taken to parasitising dunnocks and the arms race has only just begun.

'Her eggs were four of dusky hue,
Blotched brown as is the very ground,
With tinges of a purply hue
The larger ends encircling round.'

John Clare

Chapter Three
ECOLOGY IN AN EGGSHELL

Our blackbirds are now ready to lay their eggs. They have respectively sung and been serenaded, and the female has made her selection of a mate. They have displayed and courted and copulated and built a nest. In the northern cool-temperate zone, it is now about early April, and the female, being attended and fed by the male, is settling down to lay her first clutch of four eggs. Her ovary contains, in varying degrees of ripeness, several yolks, gathered together like small bunches of grapes. The most developed of these now breaks through the wall of its nurturing membrane, drops into the top of the tube – the oviduct – whose other end is the bird's cloaca, and is immediately fertilised by sperm that has been held in storage since the time of the copulations. The yolk, which now contains an embryo, moves slowly down the oviduct, gathering a coating of albumen – the egg-white – and eventually a pair of parchment-like skins. Water and salts ooze in, and the egg-white swells until it fills the skin tight. It is now ready for the shell.

Many animals lay eggs. In fact, except for mammals and the occasional 'lower' creature whose eggs hatch inside her, giving the illusion of mammal-like live birth, all animals from insects up lay them. But birds and mammals differ from all other animals in that they are warmblooded, and the newly forming warm-blooded animal must, until it has developed its own heating system, be kept warm by the parent. In all mammals except platypuses and echidnas (which lay bird-like eggs), this is done by simply keeping the developing youngster inside the mother or tight against her in a pouch. In most birds the embryo is kept warm by the parent sitting on the egg. So, unlike cold-blooded animals' eggs, which are characteristically laid in great numbers and abandoned by the mother to the mercy of the law of averages, birds' eggs have to be able to support the weight of an adult.

That is the main reason for the shell. Another reason is that the

Even without seeing the eggs in situ, *it would be possible in most cases to describe the locations of nests by the appearance of their egg-shells. The soft-coloured, dappled look of blackbird eggs (right) are the typical product of scrub-nesters, while snipe eggs (below) look very much like the straw-and-leaf matting they are laid in.*

Out of context, wood pigeon eggs (left) could not be so easily identified, because white eggs are laid almost exclusively by birds that nest in dark places. A wood pigeon has white eggs because its nest has a flimsy, see-through bottom, and from below the eggs look like the sky. Among the best camouflaged of all is the ringed plover's egg (below).

realities of external incubation mean the parent may have to leave the nest from time to time, and the eggs are then exposed and unprotected by anything but their imperviousness. This deters many potential predators, but in others it has led to some bizarre and some very clever adaptations. An egg-eating snake, for example, first dislocates its lower jaw – which any snake does when it eats anything very large – and then, without breaking the shell (indeed, *because* it can't break the shell), swallows the whole egg, which travels like a tennis ball through a hosepipe about a quarter of the snake's length until it reaches an egg compartment with a special egg-tooth. This punches through the shell; the yolk and albumen flow down to the snake's belly, and then the egg-breaking compartment folds the shell neatly and sends it back up for regurgitation.

Eggshells are very strong, but they also have to be vulnerable to the peck of their tiny inhabitants, and so they are brittle as well. This means that the successful egg-eaters are the ones that have worked out a way of, as it were, pecking in from the outside. To most birds with beaks and mammals with teeth the problem is slippery but not insoluble, but some eggshells have evolved to be very thick (and the chick inside has had to be equipped with a correspondingly powerful pecking muscle). This development stumps almost all predators except those that can use some kind of tool. Humans and the other primates can bash an egg with something – a stone or a stick or a spoon – and Egyptian vultures have learned to crack one of the hardest shells of all, the ostrich egg's, by dropping rocks on it. The dwarf mongoose crouches over an egg like a rugby player and then slings it backwards between its hind legs at a stone or a tree trunk. Skuas have been observed taking penguin eggs in their talons and dropping them from heights.

The fact that so many animals go to such lengths to get into a bird's egg is a compliment both to its nutritional value and to the quality of protection given by the shell – the egg coating which is peculiar to and universal among birds. To make the shell, the raw material the mother needs plenty of is calcium (the finished product is crystallised calcium carbonate mixed with a little organic protein matter), and if she is a blackbird it is her mate's job to see that plenty of calcium is what she gets. Female dunlins and crossbills have been observed seeking out the bones and teeth of small mammals at a time when the males are eating different things entirely. A kingfisher male, one of the birds which looks after his mate, will give her many more fish than he eats himself, and always the bigger and better ones.

The shell may be white or any of a range of colours determined first by evolutionary advantage and second by chemicals distributed from the bird's liver and mixed in with the calcium carbonate. Blackbirds' eggs range from light blue through light brown to a deep reddish-brown, and they are speckled, which is the penultimate touch before laying and which operates on much the same mechanical principle as striped toothpaste. Depending on the species and its requirements, the egg either pauses for a short time to be dappled by squirts from an array of pigment ducts, or moves and turns in the spray for smears and lines.

The main reasons for colouring and marking are, of course, to do with camouflage, and whether they are coloured or marked at all depends, in the first instance, on whether they are ever exposed to view. White eggs tend to be laid by birds that nest in holes or cavities or domed nests, such as owls, kingfishers, bee-eaters, woodpeckers, swifts (except for one species that nests in the open and lays a blue egg), petrels, shearwaters, rollers, hoopoes, martins, some warblers, wheatears, penduline tits, dippers and wrens, and by birds which are so large and/or ferocious that their eggs' shells are impossibly thick or that no predator would dare threaten them – the storks, ostriches, eagles and ospreys.

The main consideration when eggs are white is that the parents be able to see them in the dark, and to this end many eggs are not only white but glossy. At the other extreme are the eggs of birds such as the nightjar and the sandpipers, plovers and terns – birds that lay out in the open, without, in most cases, even the benefit of a nest. Finding ringed plovers' eggs on a pebbled beach or terns' eggs on a shingled one is next to impossible unless you go about it very systematically, quartering the search area and concentrating hard on a 'search image'.

The eggs of birds that use open nests are usually coloured and marked to mimic the dappling in a tree, with shadows and streaks and spots of light. The main danger to a blackbird's eggs, for example, would be a crow or another egg-eating bird flying over and spotting them, but in many cases when a bird has evolved to nest in a tree it has lost in camouflage what it has gained in inaccessibility. Wood pigeons, for example, build very sparse, almost flimsy nests with see-through bottoms, and their eggs are white. You could almost believe that wood pigeons had given up the rat-race and resigned themselves to losing

The egg of an ostrich has the thickest shell of all, but one bird, the Egyptian vulture (right), has learned to crack it open, by bombarding it with stones. The eggs of the razorbill (opposite top) and the guillemot (bottom) are defended against rolling off their precarious cliffside nests by their extremely conical shape. Guillemot eggs also have individual markings, to help parents identify them on crowded ledges.

their eggs, until you tried standing at the base of a tree underneath one of the nests and looking up through it. Unless you already knew, you couldn't tell there was a nest there, as the eggs are lost against the sky — it is white camouflage.

There is at least one other function of an egg's coloration, and that is to make it easier for the parents to identify. One reason for this might be to distinguish it from a cuckoo's egg, and a number of the markings and colours of certain songbirds' eggs are the result of an ages-long struggle against the parasites. Another reason is to enable them to tell their eggs from those of their neighbours, a particular problem for colonial sea-birds. The streaks and the marbling on a guillemot's egg, like something created in the splash-and-spill school of abstract art, are entirely random and as individual as fingerprints, as are the rather more sedate speckles on the khaki-coloured herring gull's egg. In Mexico and Central America, great-tailed grackles, which nest colonially in coconut palms, also have individually identifiable eggs, based on dappling rather than the guillemots' ground colouring.

Some gulls and terns have taken the identification function a step further and actually lay eggs for predators, usually crows, to identify. The seabirds typically lay three eggs to a clutch, and in normal times and with a normal ration of food only the first two of these can seriously be expected to hatch. If the third egg hatches that is a bonus, though the chick will be the runt; but in times of plenty, even runts survive. Nevertheless, the third egg, the one that contains the runt, is much more brightly coloured than the other two. This is so that when a crow attacks — and while doing so it is being mobbed by the gulls or terns and is, to say the least, distracted — it will go for the runt's egg first, as the most conspicuous. Thus identification, camouflage and almost every other device for the protection of eggs are turned upside-down, and at least one kind of egg is in fact designed to attract its own destruction.

The last thing that happens to an egg before it is laid — the icing, as it were — is to be given a coat of protein polish. Waterproofing is probably the main purpose of this coating, although there is vast variation in shell texture among birds, and so even the relative roughness or smoothness of an eggshell serves some survival function. Yet it is not always obvious what the exact advantage is. Why should a tinamou, for example, lay an egg of such glistening light green, blue or deep brown, and why should the eggs of flamingoes and cormorants be so chalky? It is rather easier to understand why ducks should want their eggshells to have an extra waterproof coating of oil, or why plovers would enhance their pebbled coloration with a pebbly texture – but why does an emu's egg need ridges and why does the (non-parasitic) guira cuckoo of South America need on its blue egg a relief of latticework so thick it could have been an inspiration for Wedgwood china?

There is no end to questions like this. Anyone could be forgiven for supposing there are not all that many things to know about something as simple as an egg: it has a yolk, a white and a shell, and to get at the first two you crack the third on the side of a pan. But those three elements – the yolk, the white and the shell – are about all that the 10,000 different eggs of the 10,000 species of birds have in common. And for each variation there is a definite evolutionary explanation.

Take as an example an egg's size. This varies between the .3-gram, 6.5-millimetre Cuban bee hummingbird's egg to the 5000-times-larger ostrich's egg, 203 millimetres long and 130 grams heavy. And the largest egg ever laid was the Madagascan elephant bird's, which was nearly 400 millimetres long, weighed 12,250 grams and contained more than nine litres of material, or about as much as the petrol tank on a motorcycle. It happens, rather unsurprisingly, that the bee hummingbird – which really is no bigger than the queen bumblebee you see in the very early spring and which would fit inside an ostrich's eyeball – is the smallest bird in the world; that the ostrich is the largest, and that the elephant bird, which looked a lot like an ostrich but was about a metre taller, was the biggest bird that ever lived.

However, when you notice some of the anomalies between these extremes and compare the sizes of all adults with the sizes of their eggs, you suddenly have to ask why the bee hummingbird's egg, whose weight is 15 per cent of its mother's, is so large, and the ostrich's – 2 per cent – is so small. An ostrich's egg may be 5000 times as large as a bee hummingbird's, but an adult ostrich is more than 40,000 times the size of the hummingbird adult. We already know the answer in the case of the bird with the greatest disparity, the European cuckoo's egg at only 1 per cent: it has to deceive its smaller hosts. But what about swifts, penguins and African mousebirds, which are just above the cuckoo on this scale. Why do *their* eggs have to be so relatively small? Why is it that a snipe lays an egg more than 45 millimetres long when a black-bird, almost exactly the same size, lays one that is 30 millimetres

Common gulls (right) build their nests on high, well-drained spots where they can stand look-out, and their eggs are stone-coloured for camouflage. By contrast, ostriches (below) lay on low sites in the desert sand. The eggs are white to reflect heat, and can afford to be conspicuous because the sitter always feeds close by.

The rufous hummingbird (left) provides a rare instance of white eggs being laid by a tree-nester. The camouflage is effective, with the white shell blending against the nest's cottony interior, and the interior against the general dappling and staining of the pine tree to which it is attached. Hummingbirds' eggs are the world's smallest, about 3000 times smaller than the largest, the ostrich eggs opposite.

long? Why is a kiwi's egg 20 per cent of the weight of a kiwi? Why is a crab plover's 25 per cent?

These questions are not just rhetorical. In many cases scientists are still looking for the answers, but one general rule is that a chick which hatches from a relatively large egg is relatively more developed and expects, and gets, relatively less help from its parents. Another (often broken) general rule is that chicks of ground-nesting birds – chicks that are almost instantly peripatetic and will start scratching for food as soon as their mother has led them to a suitable place, or chicks that, like kiwis, enter the world fully feathered and with about five days' rucksack of yolk to see them on their way as they immediately strike out from the nest to seek their fortune – come packaged in bigger eggs than the almost entirely helpless songbird hatchlings, immobile in their shell-like nests and still completely dependent on their parents.

While we are, as it were, rolling eggs over and examining their properties, there is one last one to consider, one that is particularly pertinent to laying: shape. Oologists, people who specialise in the study of eggs (and why not?), recognise four basic shapes, each divided into long, medium and short versions. These are spherical, the roundish eggs of owls and other birds of prey, woodpeckers, kingfishers and bee-eaters; oval, the 'egg-shaped' eggs of the thrushes, swifts and, of course, the galliformes, whose numbers include the red jungle fowl and its domestic descendant, the chicken; elliptical, the longish but more or less symmetrical eggs of hummingbirds, swifts, swallows and most of the other fast-flying birds; and pyriform, the extremely pointed, ice-cream-cone-shaped eggs characteristic of most seabirds and waders and a few ground-nesting birds such as ruffs.

The reasons for these variations are more immediately obvious than, say, the reasons for size variations. Note, for example, the coincidence between round and white, and then remember that white eggs belong to birds that nest in holes or cavities or great high-sided fortifications. White is so that the parents can see them in the dark, but white is also the absence of colour. In other words, white is the natural colour of a calcite eggshell, and if there is no *reason* for another colour to be added – no need for camouflage or personal identification – then evolution, which never does things gratuitously, would certainly not go out of its way to provide colour. The same goes for roundness. Round is the standard from which the other eggs deviate. If there is no danger of the egg rolling out of the nest, if the nest is large and warm and the parent is large and warm and there is no call for shaped 'packaging', if the aerodynamics do not require a streamlined oviduct, then nature will gravitate towards that most basic of all shapes, the sphere.

Conversely, if there is a need for any other shape, then evolution will make the required effort. The oval eggs are the 'packaged' ones, designed to fit neatly, thin ends inward, in a clutch of four, for maximum benefit of incubation in a cup-shaped nest. The elliptical eggs of the fast-flying birds owe more to the shape of the parent than to any great advantage in an egg that looks like a medicine capsule. The parent bird needs a compact abdomen for speed, and a compact abdomen makes eggs that are that shape. As for the highly pyriform eggs, they generally belong to birds that nest in precarious places and cannot afford to have eggs with any tendency to roll around. The most conical egg of all is probably the guillemot's, which is laid singly and directly on the ground in a space hardly larger in diameter than twice the length of the egg and, as often as not, the same distance from a sheer drop to the rocks and the sea. If you give a guillemot's egg a push, it will roll around like a toy top that has just fallen off its axis, and the thin end will not veer out of a circle the size of a teacup.

So, 24 hours from the time the yolk dropped into the oviduct, our hen blackbird has constructed and laid (small end first) as self-contained a life-support system as can be found anywhere on earth. It contains everything the embryo needs in the way of nutrients, minerals and water. It conducts gases, heat and light. It also conducts sound, and the hens of many species will begin schooling their chicks in their native language while they are still in the shell. It has a mechanism for waste expulsion, it has a moisture-control system and it is impervious to bacteria. In fact, there is probably no more perfect – and certainly no more compact – example of a complete ecosystem anywhere else in nature.

At the blunt end, where the egg breathes, is the air space: compare a shelled hard-boiled egg with an unbroken egg and notice that the former's blunt end is almost flat; sometimes you can stand a hard-boiled egg upright. It is the difference between that flat surface and the curved surface of the shell that allows for the exchange of oxygen and carbon dioxide between, ultimately, the developing chick and the great outside, and even when the chick has grown to the point that it can use its

Australia's cassowary, (opposite) is one of the few species in which the male does all the incubation. The kingfisher in its burrow (left) is also a male, but it shares duties with its mate. The bush stone curlew (below) also shares sitting duties with its mate.

own lungs, it pushes its beak into the air space as if into a diver's mask.

The shell is the protection and the source of calcium, the albumen is, more than anything else, the water supply, the air space is the lung, and the yolk, of course, is the food. The relative size of the yolk and the balance of its ingredients vary from species to species, but in general 'precocial' birds – the kind that have a good coating of feathers and tend to hit the ground running as soon as they are hatched – have eggs with larger yolks than do 'altricial' birds – the ones with naked, helpless chicks. At the most altricial extreme, the yolk is about a fifth of the whole egg; at the other end it is more than a third.

And lying on top of the yolk – always on top, because the yolk essentially stays in the same position whenever the whole egg is turned – is the object of the exercise, the embryo. The original fertilised cell has, in the day between conception and laying, grown to several thousand cells, but it is still no more than a tiny spot on the yolk, not quite microscopic but still not much bigger than a grain of sand.

In a few days, the body's axis is established, and the first limb buds appear, under a sort of 'chin' formation at the top of the embryo, which by now has turned on to its side and risen up slightly from the yolk. A black spot, the eye, has appeared on the proportionately huge head. The wing has all its bones in perfect formation, made from material supplied by calcium drawn from the shell, so that as the bones get stronger the encasement gets thinner and easier to break out of. Soon the yolk nutrients are flowing rapidly into the embryo's body, until eventually the yolk is gone, replaced by a chick and a sack of the liquid waste that can't be expelled through the shell. The head has moved towards the air space. The heart is working.

While the frenzy of development has been going on inside the shell, on the relatively calm outside it has been up to the parent – in the case of the blackbird and most other thrushes, and the chicken and all other galliformes, the mother only – to concentrate on the one major contribution it can make to the egg's environment: temperature control. This does not always mean keeping the egg warm; it means keeping the egg at a more or less constant temperature, usually about 35°C, and sometimes that can mean cooling the egg down. But either way, maintaining the temperature is not a matter of turning the body temperature up and down like a burner on a cooker, but of operating more like the thermostat on an old-fashioned space heater: either on all the way or off all the way. The higher the ambient temperature, the less time the mother needs to spend sitting on the eggs. A bird that spent all the time on the eggs would not so much incubate as soft-boil them, and this might go some way to explaining the wood pigeon nest's under-floor ventilation. Few eggs can survive a temperature higher than 40°C, which is only five degrees off the optimum, and a parent bird has to be rather more accurate than an old-fashioned thermostat.

On the other hand, when an egg gets very cool, often the only effect is that the chick's development is slowed down or stopped, and if it doesn't stay cool for an unreasonably long time it can still survive. Obviously, for some species cold weather is more of a problem than for others, and their eggs can remain viable in much lower temperatures. Most duck eggs tend to be fairly indestructible in this respect, and there is even a confirmed report of a mallard's egg hatching after having been frozen solid: at one point the cold had even cracked the shell.

But, if the weather is hot – so hot that the egg is perhaps in danger of going over the critical 40° – the parent has to *do* something to cool the egg. Oddly, one tactic is to cover a clutch of eggs with the bird's brood patch, exactly the same procedure as when the eggs need warming up. This, of course, uses the physical principle that when two bodies of different temperatures are put together, the heat from the warmer one will flow into the cooler one, and in this case the bird is cooler than the egg. Some sitting birds will try to keep their body heat down by opening their beaks and panting, so that, in effect, the parent is taking the heat out of the egg and blowing it into the atmosphere.

When the problem is that the sun is shining on the eggs, certain birds will spread their wings and shade them. Vultures characteristically do this, and seabirds – whose eggs often lie exposed on ledges and rocks – will dive into the water, soak their breast feathers (which they would normally try to *prevent* through oiling and preening) and then fly back to the eggs and drip on them.

Almost all birds incubate their eggs by covering them with a sparsely feathered or unfeathered brood patch in the breast, which is what a bird exposes when it ruffles its feathers before settling down on a clutch. Birds without brood patches, waterfowl mainly, will create them by plucking out their breast feathers, and sometimes birds with big webbed feet – pelicans, gannets, boobies, cormorants, darters – will incubate their eggs by standing on them. But this behaviour, again, is

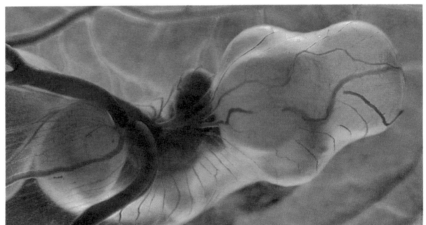

In its second day of
development (top) the
embryo of a chicken chick is
barely large enough to be
seen without a microscope,
but its head is already
discernible, as is the bud
that will grow into its feet
and wings. Two days later
(centre) it is hundreds of
times larger, and blood
vessels have appeared. By
the fifteenth day (bottom)
which is six or seven days
before it is ready to hatch, it
is a recognisable bird.

In the darkness of the Antarctic winter, male emperor penguins (top) hold their single eggs between their feet and a fold of loose skin, and get on with the world's longest incubation: 63 days. By the early light of spring, a male adélie penguin (centre) rises slightly to have a peek at his responsibility, and another devoted father, an Australian brush turkey (bottom), checks his incubator mound, a compost heap of rotting vegetation, where weeks ago his erstwhile mate buried their eggs and departed.

highly unusual. And the flightless megapodes of the Antipodes have developed the simplest all-round system of all: they search out warm compost or volcanic ash, test the temperature and, if it is right, dig a hole, lay the eggs, cover them up and go away. The babies emerge as fully feathered, independent, miniature megapodes.

Since most birds are monogamous, and are monogamous precisely so that they can cooperate in all phases of reproduction, the responsibilities of incubation are generally shared. This does not necessarily mean that parents take turns sitting on the eggs (though some do) only that they cooperate, often by a sexual division of labour. Our blackbirds, for example, behave in a way that, if translated into modern human life-styles, would probably be considered grudgingly quasi-liberated. The female mainly sits and the male sings and mainly forages. He brings food to her but not really as much as she needs, and she is often off the nest, letting the eggs cool and foraging for herself. Sometimes, if she is away longer than he seems to think she ought to be, he will sit on the eggs, even though he hasn't got a brood patch and probably isn't transferring very much heat to them. Most thrushes follow this pattern except for common wheatears and ring ouzels, which properly share the incubation.

A good general rule for guessing whether birds split incubation chores is first to ask to what extent the sexes are monomorphic: that is, how alike do they look? If it is hard to tell male from female, they probably share the tasks, and this accounts for perhaps half of all birds: almost all waterbirds, shorebirds and seabirds except ducks and geese (whose females sit while males stand sentry), storks, ospreys, buzzards (sometimes), some eagles, some vultures, most falcons, cranes, pigeons, nightjars, swifts, kingfishers, bee-eaters, woodpeckers, some larks, martins, waxbills, common starlings and about half the warblers. The species that leave incubation entirely to the female include, of course, all the lek birds and other heavily macho, visual displayers, but also some in which the difference between the sexes is not normally so discernible: owls, buntings, finches, house sparrows, thrushes and all crows except, oddly, common jays and choughs.

In about 5 per cent of species, the male is solely responsible, and in a few instances – for example red-legged partridges and rock partridges – the birds usually keep two nests going at a time, one belonging to each parent. Then, of course, there is always the longest continuous incubation of all — the sixty-three days that an emperor penguin remains completely still and patient in the worst weather on the planet, without eating, crouching and holding the single egg between the tops of its feet and its belly. And that is the male.

Meanwhile, inside the egg, the penguin embryo is as snug, warm and unfluttered as the embryo of an ostrich, or a bird-of-paradise, or an owl, or a chicken, or a blackbird at a comparable stage of development. The head, eye, body and beak are a little more than half the size they will be at hatching. All the cells have gravitated to their proper positions, and the organs, limbs, nerves and toe-nails that have been ordered by the genes to form are now at some recognisable stage of construction. For these the problem is to grow and to begin to function. For the chick as a whole, the problems are, first, to absorb food and maintain the exchange of gases and, second, to manoeuvre into an escape position.

The embryo starts from position one: it is lying on its left side with the back of its arched neck under the air space and its head tucked between its legs. In its final pre-hatching position, the chick's beak will be in the air space, and its feet will be pressing against the small end of the egg — and the next five or so days are spent getting from the one position to the other. Slowly and determinedly, like Houdini trying to get out of manacles and then out of a water tank, it twists its body, pulling its head up along its breast and straightening its legs. Finally, by way of the right underwing, the head turns the right way up and the beak pushes into the air space. All that remains is to punch through the membrane, which has been getting thinner as the water has been drawn from the albumen, and after five or six hours of twisting the head and pushing forward with the beak, the membrane breaks and the chick breathes. It claps its beak, it cheeps. Outside, the mother hears it, and clucks. Siblings in other eggs hear it, and cheep.

The chick is equipped with a special egg-tooth on the end of its beak for the one-off purpose of breaking the shell, and also with the most powerful muscle a bird ever has — the hatching muscle in its neck. Each of these is destined, respectively, to fall off and to wither once it has served its purpose. The feet are now pressed against the small end of the egg, ready to push (the chicks of large running birds such as rheas, emus and cassowaries fairly burst out of their eggs when they first straighten their knees). There may be as much as half a day left in the chick's original world as it struggles free, but the rest is only a formality. It is born.

The first-born in a coot's clutch uses its egg-tooth (top) to make the breach between the egg's air pocket and the actual outside air. Then (centre) it stretches its head to break the shell and, by straightening its legs (bottom), bursts through. Wriggling free (top opposite), it is finally able to expose its down and dry out (bottom opposite) as it awaits the delivery of its first meal.

'. . . or, with parental care,
From twig to twig their timid offspring lead;
Teach them to seize the unwary gnat, to poise
Their pinions, in short flights their strength to prove,
And venturous, trust the bosom of the air.'

Thomas Gisborne

Chapter Four
COUNTDOWN TO LAUNCH

What life looks like to a chick that has just pulled itself out of an eggshell depends mainly on whether the bird is precocial or altricial, for, by definition, it is at this stage that the two kinds are most different.

Up in the tree, the altricial chicks are, for all practical purposes, still embryos, and in fact life for most of them looks like nothing at all, since they haven't even opened their eyes yet. The sitting parent eats, throws over the side or carries away the broken shells and then settles down over the featherless brood – which still have not developed proper temperature control – just as she did when they were still inside eggs.

The precocial chick – pheasant, duckling, almost any bird that is born on the ground – enters life on its feet. It has a good coat of down, its thermostat is in order, its eyes are open, it can scratch its head, it can even preen. What it needs more than anything at that moment is to find its mother and be given instructions on what kind of bird it is, and, more important, to be shown what to eat. Its first overwhelming need is to receive an imprint.

Imprinting is a process by which, to some extent, almost all 'higher' animals establish their basic identities. It is as if the instincts are waiting, are latent, until the image of another living creature, almost always a parent, converts them into action – in a sense, plugs them in. At what point any particular animal is receptive to imprinting varies enormously, but it dates from conception rather than from birth or hatching. A human baby, for example, does not begin imprinting until much later than, say, a puppy. Likewise, a blackbird whose eyes aren't even open at hatching doesn't imprint until it is about as many days from conception as the new-hatched, active chicken chick on the ground. The amount of time it takes for an imprint to soak in also varies immensely, and here babies, puppies and other mammals are alike, because establishing what they are is not a particularly hasty process.

Birds, though, are famous for the instantaneousness of their imprint-

Once the chicks have hatched, the parents' responsibility stretches to every need they might conceivably have. Among these is sanitation, and a father bullfinch (above) regularly removes the chicks' droppings, hygienically wrapped in a gelatinous sac. From the point of view of a parent great tit (below), life's principal object is to drop food into these bottomless craws. The male blackbird (top opposite) pitches in to help his mate deliver the food, while a booted eagle (bottom opposite) not only has to deliver the meal – in this case, a snake – but butcher it into bite-size morsels.

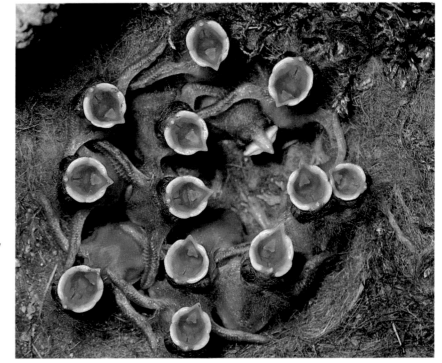

ing, and were made famous for it by the great Austrian ethologist Konrad Lorenz. By spending a few minutes in the presence of a hatchling at just the right time, Lorenz eventually had goslings trailing after him as though he were a mother goose. In another experiment, scientists worked out the 'imprint window' of two species of finch – zebra and Bengalese – and briefly 'cross-fostered' them. The chicks were switched to the wrong nests and for a little while were attended by the wrong set of parents; then they were switched back. Even though, except for that brief period, they spent all their lives in the company of their own species, when the birds grew up, the zebra finches looked for mates among the Bengalese and vice versa.

Studies of mallards have established that a chick is most susceptible to imprinting when it is precisely sixteen hours old, although, of course, it has been receiving some kind of imprint signals since the time of hatching, or before hatching if the mother has been quacking at it during the period when its head was in the air space. But the curve is very sharp between fifteen and sixteen hours, and sharper still when by eighteen hours receptiveness drops below its level at hatching. Scientists are not certain why imprinting should take place at this point in particular, but it does coincide with the time when the duckling's residual yolk has run out and when the hen would be taking her brood to water for the first time, for them to learn how to swim and, more important, how to eat.

The altricial chick, though not yet ready for imprinting and, at any rate, not in the need of an imprint to know reflexively to open its mouth, has also used up its yolk, and is just as hungry. In fact, such are the evolutionary demands on the little animal to grow at the greatest possible speed, to get itself some feathers and to leave the nest so that its parents can start another brood, that the conversion of organic material into baby bird is a 24-hour-a-day, all-consuming interest. To that end it has a voice and a mouth.

For their part, the parents are mainly but not solely concerned with responding to the voice and dropping food into the bottomless gape. They have to devote a certain amount of attention to protecting the chick from predators and weather and to general good housekeeping in the nest. In the first day or two, most of the high-nesting songbirds and other passerines will do something that seems extremely unhygienic: they will eat their offsprings' faeces. One reason is that the incubating parent is still incubating – is in fact spending more time on the nest than just before the heat-retaining eggs hatched – and, quite simply, is hungry. Besides, at that age the chicks' digestion does not work very well, and their faeces contain few true waste products but a lot of perfectly edible food. Only a couple of days later do they begin to produce true faeces, in coincidence with their yolk-sacs running out and their internal heating getting into gear. Then life takes on an altogether new pace: gathering, feeding, cleaning, looking out for danger.

Conveniently, as it happens, almost all young birds have something like built-in nappies. They defaecate inside tiny gelatinous sacs, which the parent can carry a distance from the nest and drop. In some birds, the dedication to cleanliness appears almost community-conscious: the female lyrebird takes her babies' sacs to a stream if she can find one, and actually puts them under the water. If she can't find a stream, she digs a hole and buries them.

Some passerines spread the faeces on branches, and eagles' chicks defaecate over the side of the nest. Swallows do this, too, and kingfisher chicks back up to the entrance to their nest-tunnel and squirt their waste clear – or nearly clear, for as often as not they just miss the hole, and as the brood gets older the parents increasingly become covered in faecal matter every time they go in and come out of the tunnel. Fortunately, kingfishers always live near water, and they can always have a bath after each feed, which they do. At first glance, it might look as though the kingfishers' waste disposal system has literally misfired but, on reflection, the waste *is* disposed of. It is carried in the feathers of the parent to the water: an example of a bird using its body as something like a pipe-cleaner.

Other hole-dwellers, such as hornbills and woodpeckers, have a litter of leaves on the floors of their chambers, which they change every so often, as humans change the newspaper in a canary's cage. Herons, like eagles and other large birds, do not have to worry much about droppings giving away the locations of their young, and have open grids in the bottoms of their nests. It is thought that the wet droppings landing on the branches below help cool a heron's nest – and for birds that live in warm climates, overheating of the young is a major problem, as it is for birds such as blue tits, whose nests can be very crowded and whose method of cooling is often the very simple one of not removing the faeces until they are dry. And there are, of course, a few birds that don't bother

After feeding, another preoccupation of a parent is its chicks' welfare. A newly hatched little grebe (above) finds a haven on its mother's back, while a mother red-capped plover (right) lures a predator away from her brood by pretending to have a broken wing and thus to be easy pickings. When the enemy pounces, she flies. A mother ptarmigan (opposite) provides a refuge from enemies and the elements.

with waste disposal at all; they simply put up with it until the young are fledged and then build new nests for the next brood: these include hoopoes, trogons and pigeons.

The third major concern besides feeding and sanitation is defence. And since, in general terms, defence and feeding are probably the most important two elements in shaping a species, it is probably safe – if daunting – to say that there are as many different methods of avian defence as there are kinds of food. Evolution being what it is, the most important aspect of defence is of the young – because a species in which the adult bird could protect itself but not its offspring would obviously not be long for the world.

Birds put almost all their defensive energy into protecting the individual from the time of conception through to the natural end of the reproductive phase, with most emphasis on the stage of life when the individual is most vulnerable – when it is very young. The various tactics are numberless, but can be broadly broken down into hiding from a predator, outwitting it, outrunning it or successfully attacking it.

The first category includes all the permutations of camouflage and is without doubt the most common. All birds hide to some extent; even eagles and vultures build their nests in inaccessible places, which might be considered a form of hiding. Perhaps the least subtle are some of the colonial seabirds, whose eggs and young seem to lie around like so many gifts for overflying skuas and gulls, but it has to be remembered that these birds of prey are the seabirds' *only* predators, that they have made themselves absolutely inaccessible to anything with fur and four legs, and that the population balance is still heavily in favour of the prey. In other words, skuas and gulls can eat their fill, and there are still plenty of seabird eggs and young – and the chances of any one seabird losing one of its gene-carriers remains pretty low, about fifty to one. They don't need to hide.

As for outwitting, perhaps the best-known tactic is the partridge's broken-wing ruse. This is when the ground-nesting hen hears approaching danger and leaves the nest before the predator sees it, and she drags her wing across some exposed piece of ground. The fox, or whatever, duly gives chase and suddenly discovers, as she flies away, that the wing wasn't so bad after all. By then, the predator is a long way from the nest, and has little hope of ever finding it.

Almost all plovers use this tactic, too. Other species will sometimes crouch low and scurry through the grass, squeaking in imitation of a mouse, only to spring into flight when the fox or cat is about to pounce. Some of the galliformes pretend to be casually pecking and scratching in some place away from the nest, the hen apparently oblivious to the predator that she is in fact perfectly well aware of, and others will sit on a nest where there isn't a nest. There are several versions of this kind of distraction, some involving 'wounded' flying, some involving sudden movements, others involving mock nest-building or mock sleep. Perhaps the least subtle form is mobbing, when a lot of small birds, not always of the same species, will attack a stoat or a fox or a bird of prey, not in any hope of injuring the larger animal, but of taking its mind off its immediate goal.

One very specialised ruse is the one used by the cape penduline tit, which builds into its hanging nest a decoy entrance and chamber, so that potential predators – in this case usually snakes – will go in, find it empty and presume that another snake has got there first. Meanwhile, the chicks are alive and well in another compartment entirely, reached by a camouflaged doorway that actually opens and closes.

Birds are often forced to be clever by the fact that they are so physically specialised. All their bodily contours, limb capabilities and other characteristics are geared towards one supreme talent: usually flight, but also swimming or ground speed. None of these makes much allowance for the ability to pick up and carry anything beyond light, graspable loads that can be handled by a beak. There are nevertheless a few birds that have developed ways of grabbing the chicks and getting away. One of these, the American sun grebe – an odd waterbird that looks like a cross between a grebe, a coot and a cormorant and hides itself from predators by swimming with only its head sticking out of the water like a periscope – has, under each wing, special pouches into which the chicks have been tucked since soon after hatching. And there they stay until they can feed and otherwise fend for themselves.

Birds without this kind of physical adaptation (and as far as is known the sun grebe is the only one with it) can still sometimes manage to flee from predators while carrying their young with them. Some parent waterfowl call to the chicks to jump on their backs before they swim away from a land-bound mammal. Woodcocks can fly while carrying chicks between legs and body, and red-tailed hawks will fly away with

Two species which give their young milk — very much like mammalian milk except that it is produced in the bird's crop and dispensed through the beak — are greater flamingoes (opposite) and emperor penguins (left). Both kinds of milk have claims to fame: a penguin's is the only milk in the world produced by a male, and a flamingo's is the only milk that is red.

their young in their talons. Rails actually carry them in their beaks.

But, in general, the ability to carry is not a prime avian talent. Ingenuity is. And, as with deception, it is exercised quite well when a bird has to resort to threatening or attacking something that is threatening or attacking it. Sometimes the young themselves are specially equipped: hoopoe chicks can emit a vile odour, and baby fulmars will spray a rancid-smelling oil on anything that starts to show an interest in their nest. The chicks of birds that nest in holes – owlets are among the most adept at this – can usually produce a chilling imitation of a hissing snake, which is enough to deter almost any intruder *but* a snake. Gulls build a secondary nest, a sort of bomb-shelter, for their chicks to retreat to, and while that is happening, the parents and other members of the colony will stage a round of screaming aerial attacks on the unfortunate intruder, plastering it with faeces. Even that otherwise mild-mannered thrush, the fieldfare, has been observed to attack small birds of prey by bombarding them with excrement.

As heroic as some of these parental efforts may be, to the chick they are as irrelevant as a parent's professional occupation is to a human baby. The overriding consideration now is food. In fact, there is a certain conflict of interests at this stage, and it is possible that the constant noise altricial chicks make is a sort of blackmail, playing on the parents' fear of predators: they feed chicks faster in order to shut them up.

The exact ways in which the parents meet their offspring's demands are too various to enumerate but, broadly speaking, chicks are given more or less the same food that the adults themselves eat but in different proportions, with more protein for growing babies. Birds that normally eat a mixture of plant and animal – usually invertebrate – food will drastically increase the insect content for the nestlings, to provide the extra protein.

Some birds have a built-in vitamin, mineral and protein control system, in that they have a sac attached to their oesophagus, called the crop, which in most birds is used for food storage. The crop produces a liquid that is so much like the milk mammals produce that there is no good reason not to go ahead and call it milk – crop milk. All the members of the pigeon family have this capability, as do greater flamingoes and emperor penguins. Pigeon milk is 74 per cent water – an important ingredient to a nest-bound chick – and the rest is 58 per cent protein, 35 per cent fat and 7 per cent minerals, which is a good deal richer than the milk of any domestic animal used by humans as a milk source, and for that matter a good deal richer than human milk.

The mother pigeon produces the milk for fifteen or sixteen days after the chicks have hatched and feeds them by regurgitating it into their open mouths. As the fortnight progresses, the young are gradually introduced to solid food until they are weaned. For parent pigeons it means that they can eat whatever comes and break it all down into the correct chemical balance without having to go very far from the nest. It also means that they can incubate their chicks and feed them at the same time, thus overcoming a problem that is, for almost all other birds, one of life's principal preoccupations.

Flamingo's milk has about the same amount of water as pigeon's, but about twice as much fat and half as much protein, and it is all a flamingo chick eats for its first two months of life. And it has one characteristic that makes it different from all other milks in the world – it contains the pigment canthaxanthin: it is red. And emperor penguin milk has its own claim to fame: it is the world's only milk produced exclusively by the male of the species. After incubating for sixty-three days, the male may find that his mate is late in returning. If so, for a few days he can feed the chick on a special secretion.

Other variations on the two-parent, constant-activity feeding system include communal feeding, when close relatives, often siblings from the year's earlier broods, will help out. About 30 per cent of a second- or third-brood Florida scrub jay chick's food is brought to it by elder brothers and sisters. In house martins, not only do earlier chicks from the same nest help, but earlier chicks from other nests. But one bird, the pukeko, a New Zealand moorhen, has taken this tendency so far that it has revoked the usual one-sided relationship between a clutch's biggest chick and the runt. If one chick starts falling behind, the bigger ones will fatten it up.

By and large, though, the feeding arrangements for altricial birds are not much different from the ones used by our blackbirds. The chicks are born blind, naked, cold and with their mouths open, and the parents respond to an open mouth by putting food (in the case of blackbirds, mainly chopped earthworms) into it. The mother is the main feeder, incubator and defender, and devotes all her available time to it, eating for herself the smaller bits she forages in the field and, for reasons of carrying economy, taking the bigger pieces of food back to the nest. The

As chicks approach fledging day they get more curious about the world beyond the nest, or, in the case of a young emperor penguin (above), beyond mother's brood patch. Tawny owlets (opposite) will sometimes venture outside their native tree hollow. When their eyes begin to open, Alaskan dippers (above left) will learn to close their gapes from time to time in order to look around, as will the chick of a white-tailed tropicbird (left).

male labours as he is needed. If the weather is unseasonably cool or rainy and the hen feels obliged to concentrate on incubation, then the male forages and carries. If it is still and summery and the pickings are easy, the male will be more concerned with singing and preening, foraging for himself and looking for a place to have a bath.

In the meantime, the birds of prey chicks are having lessons in what and how to eat, and how to look out for danger. In the case of eagles, the food is being brought to the nest in the form of whole rabbits or mice or pigeons, to be torn up there by the parent into small chunks which are sometimes not small enough, so that the chicks will have to cut it up a bit more, to learn how. And on the ground, in the grass, at the waterside, the precocial chicks are trailing around after their mother, eating, like her, what they find by scratching or diving or giving chase, trying not to get lost, cheeping, running to the brood patch at every sign of danger. And all the time they are growing feathers.

One of the oldest and most hackneyed of the arguments against the theory of evolution – and one that was used to great effect in the days before there were many details to back up the scanty palaeontological evidence of the early birds – was that there was no logical way a bird could have developed feathers and wings from, as it were, a standstill. That is, a wing either works or doesn't work, and there could have been no impetus towards development of a wing unless you gave the whole process of evolution something like foresight – a thought which was anathema to evolutionists, and sacrilege to their opponents. The best the Darwinists could say then was that observation as yet was incomplete, that somehow the wings and feathers were developed for another purpose, and that one creature or another finally discovered that it could spread its forelimbs and glide for short distances, giving it a slight speed advantage on its prey.

Most of the scientific history of the past century and a half could be written around an argument such as this, and the evolutionary viewpoint would still be hampered by birds' perverse refusal to fossilise. By necessity, birds have light, hollow bones which decompose before the required geological principles can get to work on them. But, since the time of Darwin, there has always been *Archaeopteryx lithographica*, Latin for 'ancient wings which have left their image impressed in stone'. First found in 1861, the particular slate was excavated in a Bavarian quarry, and the particular fossil, which was handed over by curious quarrymen to a local museum, differed from all the other *lithographica* fossils turning up in those days, by showing the first clear outline ever seen of a prehistoric feather. The rest of the creature was there, too, and it was possible to reconstruct, from 150 million years ago, the earliest known bird – or one of the most advanced of the dinosaurs. It was not much larger than a pigeon, had stumpy but very definite wings that retained reptilian claws on their forward surface, a long and lizard-like tail fringed by feathers, a mane of down along the back of its neck, and toothed jaws in a long, scaly face. But perhaps the most telling piece of its anatomy was its breastbone. This was very small in comparison with a modern bird's. It was no bigger than a lizard's and could not have supported flapping wings. And its tail did not have a bird's rigid base. Archaeopteryx could not fly.

So what was it doing with wings and feathers? The theory now (and the wonderful thing about evolution as a theory based in logic rather than blind faith is that it is so eminently adjustable) is that the feathers evolved not for flight but for insulation, as a by-product of the evolution of warm-bloodedness. Certainly they are made of the same raw material – keratin – as fur, and the prime function of fur has never been questioned, because only rarely has it developed any important secondary functions (porcupine quills, etc.). Keratin is also what reptile scales are made of, but reptile scales, for no better reason than they don't need to be, are not good insulators. So it seems straightforward enough to say that as the gradual development of internal heating proceeded, the scales of what at the time was the earth's dominant life-form developed into insulating fur and insulating feathers in the two new reptile-derivatives.

The property of an insulator which makes it good at its job is its ability to hold a layer of air (or water) between the thing that is insulated and the cold outside world, and on the whole, feathers are better at this than fur. The latter is almost always supplemented by a layer of fat in very cold places, while feathers usually have to do the job by themselves. In most land-bound mammals, weight is not *that* important a consideration. With the sort of light, quick, tree-dwelling, warm-blooded lizard that Archaeopteryx had become, a layer of fat would have been a decided encumbrance. Better that the keratin coating should develop an even greater degree of airtightness – insulation without much weight.

The factor that makes evolution possible is lots and lots of time, and,

Growing feathers and learning to take care of them is the chick's principal task. The young tropical pigeon (right), in spreading its wing to the rain, is learning to bathe. For adults, feather-care involves bathing, dust-bathing, airing and preening with oil from a gland at the base of the tail. This is one of life's main preoccupations, as demonstrated by the black oystercatcher (above) and the American anhinga (opposite).

given this, it is not particularly amazing that some of the light tree-dwellers should find another use for their insulation's airtight qualities. Strengthened slightly and extended, it would not only hold air but it would allow the animal, Archaeopteryx, to ride upon it for short distances. Thus was established the principle of gliding as a first step to flight, an ability theretofore confined to the highly successful insects, and it was such an advantage that it would inevitably lead to a new class as great as or greater than even the reptiles.

Feathers themselves now exist, on any one bird, in four main forms: the down, which is still an insulator; contour feathers along the breast and back for streamlining; sensory feathers, single shafts around the eyes and sometimes the base of the bill that serve the approximate function of a cat's whiskers; and flight feathers. And all of them have become highly sophisticated tools which need constant maintenance through oil preening and water bathing, and even dust bathing.

Secondary uses of feathers are camouflage and display. Primary uses can also be either dispensed with or converted into something else. Flightless birds, obviously, do not have any great need for flight feathers, but some still have them to give them lift in their main means of locomotion, be it running or swimming. A fast-running bird such as an ostrich or a rhea, for example, will hold its stubby wings out and to the back as its speed increases. And the underwater motions of many swimming birds — penguins in particular — look very much like aerial flight.

It should always be remembered that flightless birds are not dinosaurs which never got off the ground. They actually descended from birds that could fly. Their ancestors returned to the earth or the water for a welter of environmental reasons adding up to the fact that things just worked out better that way.

The germ of a feather is known as a papilla — a little ring of cells on the skin of the fledging chick, or, indeed, on an adult that is having a replacement feather constructed. These cells' genes have told them that the dictates of their 'basic property' now require them to form a tube, which, as it grows, is thick and strong on the north and south of the tube's circle of diameter, and weak in the east and west. From the weak sides of this shaft spring barbs, individual hairs, and from the barbs come notched barbules, which interlock with the barbules from the barbs above and below, creating a weave that is virtually impermeable to air.

Because every piece of the structure is hollow, it is as light as keratin can possibly make itself. Then there is one last, unusual step. When the cells have multiplied and created the hard structure, there is no longer any need for living matter. The feather dies, which makes it even lighter.

Flight feathers grow only on a bird's wings and tail. If a wing is spread out, at the anterior end are shingles of contour feathers known as wing coverts, which act as sheaths for the shafts of the flight feathers. These grow towards the back of the wing and are either what are known as 'secondaries' – the relatively immobile small feathers at the body end of the wing – or 'primaries' – the big feathers at the end of the wing, the feathers that control lift and direction, the feathers, indeed, that control flight.

The other flight feathers, the ones in the tail, are the stabilisers and rudders. The transformation of Archaeopteryx-type creatures into true fliers owes as much as anything to the development of a sturdy tail, with all feathers fixed firmly in the base of the pelvis. The tail feathers are the very last bits of anatomy that a fledgling acquires, as though they were the bird's crowning achievement.

An idea now largely discredited, but which still remains attractive, is that an animal in the process of growth from conception to adulthood passes through all its previous evolutionary stages. Whatever its validity, this concept is still good for purposes of illustration, for we have seen the chick pass from a single cell, to assorted colonies of cells, to a barely differentiated insect-like, fish-like, tadpole-like creature, aquatic, air-breathing, eating, seeing, calling, fledging. All that remains is to fly.

When the time comes, the parents stop feeding the chicks or place food out of their reach. Finally, reluctantly, a tree-nesting chick will jump and flutter to the ground. If it is one of our blackbirds, it still doesn't have tail-feathers. It is fat for its size and ungainly, and doesn't yet know how to feed itself. For a couple of days it is still bound by gravity, disoriented and vulnerable. Generally it will find a perch in some shrubbery and wait for its parents to bring it food. Then one day they stop doing even that. At the same time it discovers it has tail-feathers, and perhaps sees other blackbirds feeding. It goes to them. The season is summer, and it is a bird – and instinct and evolution soon take over.

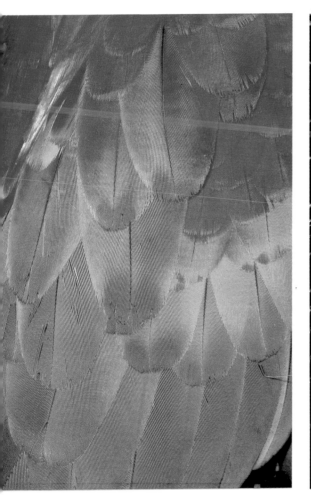

Feathers are used not only for flight, but for warmth, waterproofing, display and identification, and there are nearly as many different patterns and colours as species of birds. This selection includes (clockwise from above) scarlet macaw, peacock, vulturine guinea fowl, varied thrush, mallard, germain peacock-pheasant, superb bird-of-paradise, and Amazon parrot.

'A bird came down the walk
He did not know I saw
He bit an angleworm in halves
And ate the fellow raw.'

Emily Dickinson

Chapter Five
CRACKING THE NUT

It is not clear exactly when it first dawned upon the people of Mauritius, and upon the steady trickle of biologists with a fascination for the strange wildlife of this isolated island, that a long, long time had passed since anyone had seen a sapling of the tambalacoque tree. But, eventually, as the remaining trees began reaching the end of their natural 300–500-year lifespan, the phenomenon passed from a curiosity to a conservation problem of some urgency.

By the late 1970s there were only about ten tambalacoques left. They seemed to be healthy, if old, and were still flowering. They were being cross-pollinated (scientists were sure of this because they themselves were acting as bees) and were producing seeds. But there it all stopped. The hard-shelled, golf-ball-sized seeds could not be persuaded by any botanical technique, age-old or modern, to germinate and grow. The species had thrived in this little patch of the earth for millions of years, and though its environment had changed with the arrival of humans in the seventeenth century it could not, as far as the tree was concerned, have changed *that* much. Some of its forest was still undisturbed – the soil was the same, the climate was the same – so why would the seeds of a particular species suddenly give up? It was almost as if its controlling genes had decided to commit suicide.

Of course, genes no more do that than marbles roll uphill. There had to be a reason, and it almost certainly had to be an ecological one. The youngest tree was about 300 years old, and in 1979 an American botanist asked himself this simple question: what happened in Mauritius 300 years ago? Well, lots of things, mostly to do with the early human settlement of the island and with its establishment as a trading centre. But the most famous occurrence of all, also connected with human settlement and the devastating human introduction of monkeys, pigs and cats, was the extinction of Mauritius's best-known bird, the dodo.

A bird's beak is its eating implement, and the shape is adapted to deal with the principal ingredients of its owner's diet. The tough, stumpy beak of the rock ptarmigan (left) is used mainly for plucking buds and shoots, but the long, delicate beaks of the snowy egret (opposite) and the yellow-billed stork (below left) are used for grasping fish.

What, the scientist asked next, do we know about the habits of the dodo? The answer to that was: practically nothing. Seventeenth-century adventurers and colonists were much too busy conquering nature to sit down and observe any of it, and so information about the dodo has had to be deduced from bones and other mementoes, just as if it had gone extinct in the Pleistocene period. Still, it had long been established that the bird was a Columbiforme, a pigeon, and pigeons eat mainly seeds, fruits and berries when in season, and also flowers and young leaves. And the one dodo characteristic that the sailors and settlers did notice and note, when they were preparing the bird for the pot, was that it always had a large stone in its gizzard.

This, the biologist reasoned, would have given the dodo prodigious grinding powers, perhaps powerful enough to crack the tambalacoque nut. Turkeys, too, have exceptionally strong gizzards, and so he fed turkeys some of the seeds, collected them at the other end, planted them and saved the tambalacoque tree. The turkey gizzards had softened the casing enough to allow a seedling, for the first time since the seventeenth century, to break through. The dodo and the tambalacoque, it appeared, had evolved in a constant competition in which the bird would continually increase the strength of its gizzard in response to the tree constantly toughening its seed-cases. For this kind of stand-off to work to the benefit of both species, it has to be very finely balanced. The dodo would have had to be able to open and digest a certain number of the seeds or it would have stopped eating them, but some of the casings had to be hard enough to survive the gizzard's battering so that they would be passed intact, if much softened. In fact, the relationship, though balanced, was not equal. The dodo could have eaten any of a number of things. The tree, in developing a defence against the dodo, created a seed-casing that was so hard that it *had* to go through the bird's gizzard. The tree was hooked on the bird and would have followed it into oblivion if it hadn't been for the timely intervention, by way of apology for having killed the bird off, of a human's deductive reasoning. If the tree had gone extinct first, the dodo would have missed one of its favourite seeds for a while but otherwise would hardly have noticed.

All living things have to eat, and the only things they can eat are, or are produced by, other living things. And with the exception of fruit and some nuts – including, in a way, the tambalacoque nut – no living thing 'wants' to be eaten. It develops defences. The eaters overcome these. More defences, some better than others. Sometimes a treaty is struck, and animals agree on an occasionally uneasy symbiosis – and then gang up on something else. Sometimes one creature can take nutrients from another without it knowing, as many bacteria do when they live inside the bodies of larger animals without harming them. Sometimes the larger animals do know, and fall ill. Certainly the smaller creatures such as insects know when they are being preyed upon by the larger ones, and some of the cleverest defences of all are the ones they devise – tricks of chemicals, colouring, armour and attack.

But no defences are universally perfect, and it's a good thing, too, because it means that every living creature gets something to eat. The dynamic that keeps the living world going is the constant development of defences and attacks, none of which ever works all the time.

As for birds, a class containing 10,000 species, nearly anything that is organic and is in a place where a bird can reach it is eaten by some bird at some point. To start at the 'bottom' of the notional evolutionary class system – with the lower plants, the non-flowering ones – we have New Guinea's pygmy owl, which lives on fungus, and Africa's lesser flamingoes, whose heads hang upside-down on to the surface of warm lakes while the upside-down beaks filter out microscopic blue-green algae, the smallest food particles taken by any bird. Seaweed is eaten by wigeons, which on Arctic summer days fan out across the shallows of the estuaries like so many savannah ungulates. In the same estuaries, Brent geese eat eelgrass, an estuarine flowering plant, and on the neighbouring hillsides ptarmigans are subsisting on the shoots or buds of lowly alpine plants. The spruce grouse of both Siberia and North America eat the needles of their namesake, as does the largest grouse of all, the Eurasian capercaillie.

Generally, it is a plant's roots, flowers, fruits or seeds that birds go for, and not the leaves. This is probably because most leaves, including grass, are so interlaced with cellulose that an animal needs a special apparatus – a rumen, usually – to digest them, and considerations of airworthiness mean that it is altogether better to eat normally digestible food and dispense with the extra weight in the fuselage. Cows, sheep and other animals that do have rumens and can graze and browse are not normally among nature's lightweights.

In fact there *is* one bird which has very recently been discovered to have a rumen: that strange throwback, the Amazonian hoatzin, the

South America's hoatzin (top left and right) is one of the few birds on earth adapted to eating leaves. Not only does it have a beak like a pair of scissors, but it is the only bird that has a rumen, a feature otherwise known only in mammalian herbivores. The masked flower-piercer (left centre) 'robs' flowers of their nectar by going in through the side instead of around to the front, as the 'proper' rufous hummingbird (right centre) does. The hawfinch (below) has a beak adapted to crushing nuts, but it also eats berries.

The little owl (left) normally eats insects, but it is also equipped to handle small mammals and birds, as well as the occasional reptile or amphibian. This one has found a frog.

species whose young have Archaeopteryx-like claws on their wings to allow them to climb up and down their riverbank trees, to get to the water, where they dive and do a sort of breast-stroke. All these talents are lost when, after a very long time in bird terms, they fledge and become adults. It is almost as if hoatzins have a two-phase life-cycle, like a moth. The adult has no claws, cannot swim and is very clumsy. It does not fly well, and tends rather to leap from branch to branch, desperately balancing itself with its wings, because among the other things it doesn't have is a good perching grasp – and it carries this oversized rumen at the top of its breast, where other birds have crops, and thus has a constant tendency to topple forward. Examination of its cells suggests that it is an aberrant relative of the cuckoo, but if it looks like any other birds at all, it would have to be the galliformes.

And in the midst of all this primitiveness it has one of the most recent of avian developments, the songbird's syrinx, which it uses to make extremely loud, gravelly cries and searing hisses. 'Hoatzin' is said to be an Amerindian word for the sound it makes, but another name for the creature translates as 'stinking bird'. Never mind – it eats leaves more effectively than any other bird on earth.

Some pigeons also eat leaves, and in New Zealand, the land without native mammals, the kakapo is thought to have filled the niche occupied elsewhere by rabbits, because it grazes in the same way. A few birds, such as the palm chat of Hispaniola in the Caribbean, eat flower petals, and some others, most notably the members of the tanager family known as flower-piercers, rob flowers of nectar. Yes, 'rob' rather than simply sip, because, unlike hummingbirds, their pointed beaks punch through the side of the flower, killing it and taking its nectar without coming into contact with any of its sexual parts. The poor plant lives up to its side of the bargain, which is to produce sweet, nutritious nectar, and only asks that the bird come and get it through the front of the flower, so that when it visits the next flower of the same species it will be carrying some pollen – but the tanager cheats. Of course, the individual bird, because of its genetic programming and other factors such as the shape of its bill, has no choice in its method of nectar-gathering. The evolutionary strategy, though, is still a cheat.

So are most evolutionary strategies. Life on earth is strangely free of any moral imperative, and some of the most successful organisms are those that have devised ways of getting others to do the work for them.

All degrees of parasitism up to and including symbiosis – and including the civilised human practice of taking advantage of other animals' and plants' instincts to turn them into domestic food factories – are forms of cheating. Yet flower-piercing does seem more unfair than most, if only because it would seem to take no extra energy to go around to the front of the flower, to pay for the nectar. But the reason it doesn't is, in a sense, the plant's own fault. The flowers have evolved to attract, among other nectar-sippers, hummingbirds – and hummingbirds are about the largest of the animals that play by the rules, because anything larger couldn't afford to spend the energy required to fly up to the flower, hover and sip for the small amount of nectar it would get in exchange. And the amount is small because it is not in the flower's interest to lose all its nectar for every single hummingbird visit. It needs to spread its pollen more broadly than that, and so it works out the minimum amount of nectar a hummingbird needs to cover its energy investment and make a small profit, and holds the rest back. A flower-piercer is simply saying, to hell with that, I'll go straight to the honeypot. The only way the plant has of countering this strategy is to make more flowers than it would otherwise, which, until they are killed by the piercers, means more for the hummingbirds.

Many more birds take its fruit and seeds than take any other part of a plant. When an animal eats a piece of fruit it is partaking of the few sources of nutrition in the biological world that are 'meant' to be eaten. The other parts of the plant all have separate functions, and the plant would 'prefer' that its roots, stems, leaves, petals and nuts be left to gather and transport nutrients and water, to convert sunlight into food, to advertise nectar and to germinate and grow into a new plant. But the plant manufactures fruit specifically for an animal to eat, in the hope that it will also swallow the seeds and pass them in some other part of the woods.

It is also in the plant's interest that its customers remain healthy and come back for more, and to that end it stocks the fruit with as many nutrients as it can. The animals can tell that the nutrients are there by their taste, and many very disparate animals have evolved to agree that the 'best' taste is that of natural carbohydrates: sweet. Birds which eat fruit are nature's easy livers. Instead of competing with their main food source in a sort of arms race, they are cooperating with it, and their only argument is with other fruit-eaters.

Most hornbills eat fruit; certainly that is what their beaks are built for. But it sometimes happens that animals will contradict their basic design and assume a life-style to which, by all appearances, they are unsuited. The main diet of this yellow-billed hornbill (opposite) is insects. More conventional is the masked parrot (left) which, with a beak built for gouging and a tongue for holding, has a diet of fruit occasionally supplemented by nuts and roots.

A resident mistle thrush is a fierce defender of a tree of berries during the annual invasion of redwings — another kind of berry-eating thrush — and can chase away dozens of them. Almost all the parrots are fruit-eaters, and, in comparison with other birds, they have lives that seem almost indolent, with lots of loud calling and squabbling, no inkling of a necessity for stealth and very little need for any parrot to leave its grove of tropical trees, many of which have agreed on a sort of yearly schedule. Each tree has its allotted time to come into fruit, so that there is as little competition as possible for the parrots' attentions, and the parrots have one kind of fruit or another all year round.

The hornbills of Africa and Asia and their New World counterparts, the toucans — all of which eat the larger fruits — have long bills so that they can reach the tropical wild figs growing at the ends of branches too slender to take their weight. The oilbird of north-eastern South America is the only nocturnal frugivore, working the trees at night while the parrots and others are asleep. Because of this, it is one of the few birds of any kind to have developed a good sense of smell.

In fact, one of the reasons often given for the disproportionate affection that humans seem to have for birds, in comparison with the more closely related mammals, is that birds and people share the same dominant senses — sight and hearing — and see and hear things by much the same kind of mechanism and in much the same way. While most mammals have a very narrow range of colour vision, primates such as monkeys and apes, including man, see colours as birds see them. Birds also tend to inhabit the daylight and to make themselves visible, and they create sounds that people can hear and, to some extent, appreciate, if also usually misapprehend.

If birds in general behaved like the oilbird, sniffing around at night and stealing fruit from its daytime 'owners', they would probably be regarded with a fair amount of loathing and terror, like the poor bats. As it is, owls are imbued with a certain undeserved psychological depth, even wisdom, and nightingales, though they may decide to sing at any point in a 24-hour period, are remembered the best for the song whose beauty is made slightly sinister for having come out of the darkness.

The few other strictly nocturnal birds — some of the petrels; the little penguin; the swallow-tailed gull; the kiwis, which in the absence of nocturnal land mammals in their native New Zealand occupy their niche, as does the kakapo; and the bat-hawk, which lives in Africa and

Asia and feeds on bats – are not much in people's awareness and thus not given a chance to become the stuff of ghost stories. The oilbird, though, seems to go out of its way to be spooky. On the odd occasion when it is actually seen, it looks like a cross between a hawk and a nightjar, with a low, almost hunched posture and a bewhiskered, curved beak. As if it weren't enough that it occupies the night, it nests in colonies in caves. In this total darkness, it finds its way around by echolocation – sonar – which involves a relentless clicking sound, and when the birds are roosting they snarl and scream at each other like banshees. Outside at night they utter the occasional isolated harsh shriek as they descend on the palms, snatching the fruit in their beaks and taking it back to the cave. In Trinidad the bird is called 'diablotin', little devil.

But the fruit is not the part of the plant most commonly eaten by birds. This is the seed. For a start, all of the members of the most numerous bird family – the finches, with 400 species found everywhere but the polar regions – are seed-eaters. They include all of the birds to which the name 'finch' has been suffixed – chaffinch, bullfinch, hawfinch – as well as buntings, crossbills, canaries, sparrows and many others. They all have in common small, strong, conical bills, strong jaws and strong skulls to go with them, not to mention very strong gizzards. They do not all eat seeds exclusively, and certainly do not all eat the same kinds of seeds in the same way, but they are all built for basic nutcracking.

Every finch has a groove in the side of its mouth for holding a seed while the lower jaw crushes it. The bird then peels off the husk with its tongue, spits it out and swallows the kernel, to be ground down by the gizzard. The various species are variously equipped to handle different seeds of different sizes and toughness of husk. Tiny goldfinches and siskins use their tiny scissor-like beaks to pry out thistle seeds, and hawfinches, with bills like pliers, can open some of the larger tree nuts. The odd-shaped beak that gave the common name to the crossbill is used for extracting seeds from pine cones, and the pine grosbeak has a rounded beak which has evolved out of seed-eating altogether and into the business of crushing buds. Many finches supplement their diet – or, more usually, their offspring's – with insects, but a large group, including goldfinches, linnets, serins, redpolls, rosefinches and cardinals, never venture beyond an all-seed menu and feed their young on a sort of seed porridge which they mix in their gullet and regurgitate.

The crossbill (above) has the typical finch's beak, except for the variation that has given the bird its name. This helps it prise open pine cones. The song thrush (below) is the only member of its family that regularly eats snails, having developed the technique of smashing their shells on rocks.

Two species of macaw (top) – scarlets, with the yellow in the wings, and militaries, with the green in the wings – flock on a cliffside salt lick. Scarlet ibises (centre) also feed in large flocks. which fan out over the tropical American wetlands in pursuit of molluscs, aquatic insects, fish and frogs. The black heron (bottom) is a feathered fisherman, but its hunting technique includes drawing its wings up like a cloak, to lure fish into the shade provided.

Finches are the family of birds best adapted to eating seeds, but other kinds of birds will eat them in addition to their main fare. Parrots and their smaller relatives, the lories, parakeets and budgerigars, will turn to seeds at times of year when fruit may be scarce, and any of the tits, which prefer insects, will often eat seeds in the winter, holding them with the foot and hammering them with the bill. Large macaws can crack the nuts of the oil-palm, as can the strangest of the vultures, the palm-nut vulture. Though it sometimes eats the carcasses of fish and other creatures that have washed up on African beaches, it eats the huge nut as well. In fact this is its main diet and thus it qualifies as the only vegetarian bird of prey.

On the other hand, there are very many carnivorous birds that are not birds of prey. In fact, there are more species of carnivores than herbivores — if the omnivores, which eat everything, are classed with the former. The avian hunting habit that comes first to mind for most people is the thrush scanning the ground in pursuit of that living length of pure nutrition, the earthworm. To our young blackbird, the first lesson in life will be to imitate the adults' method of systematically searching the feeding ground, moving across it as if quartering it for a survey, looking for signs of a worm, rapidly stabbing the ground until worm comes to beak, and then squeezing it, pulling it and pecking it until it gives up the struggle.

When a bird becomes a hunter, rather than just a gatherer — which tends to take place under cover, exploiting a food source at a relatively predictable location — it exposes itself more to *being* hunted, and so the pursuit of the earthworm is interrupted regularly for a glance around or at the sky. The blackbird will drop the worm in the middle of subduing it and, as the prey dazedly tries to re-enter the soft ground, will cock an eye towards the sky before returning to business.

Many birds — and sparrows are the easiest to observe doing this — forage in flocks, to minimise the time spent on vigilance, as one or two birds can watch out on behalf of everyone. This is not to say that there are one or two birds assigned to sentry duty in a flock of, say, thirty, but that each of the thirty birds is aware of the flock's size and of the timing of the other birds' skyward glances, and so cuts its own glances by a factor of fifteen or so and somehow knows when its turn comes.

Another advantage of a flock (or a herd or a school of fish) is that it diminishes each individual's chance of being picked off. A sparrowhawk can see one or two sparrows almost as easily as it can see forty, but it can only take one at a time. Safety in numbers, for any creature, usually means getting high odds in a lottery that it is undesirable to win.

The thrushes, the omnivorous finches and many other birds will take a sizeable risk in exchange for an earthworm, and gradually lesser ones for the insects and other small creatures that may be less nutritious. One prize that only the song thrush, of the thrushes, seems very good at exploiting is a snail, which it cracks against a rock or some other 'anvil', extracts from the broken shell by bouncing it on the ground, cleans by wiping it on the grass, and then eats.

In fact, the idea of using an anvil is something that seems to pop up in isolated instances throughout a wide range of unrelated species. Australian bowerbirds do it, as well as Australian mud-nesters, jewel-thrushes (bright tropical birds of South-east Asia and Africa, and no kin to the true thrushes) and one of the kingfishers. Crows will drop molluscs, and there have been reports in England of them even dropping walnuts. Lammergeiers, or bearded vultures, drop bones on to stony cliff-edges, both to get at the marrow and to smash the actual bone into manageable fragments, and tortoises have been seen getting the same treatment from golden eagles. Without stretching the imagination too far, the anvil principle can be seen as rudimentary use of tools. Perhaps the hammer principle, as employed by the Egyptian vulture on the ostrich egg, is a clearer demonstration. And there is a recorded case in Israel in 1980 of an Egyptian vulture swooping down on a monitor lizard, grabbing it in its beak, carrying it to a height of about 100 metres and dropping it. When the vulture flew back down to its victim, the lizard was still alive, and so the bird picked up a stone and beat it on the head until it stopped moving. In Australia, white-winged choughs have been seen using broken mussel shells to hammer open unbroken ones, and, also in Australia, brush turkeys will kick barrages of stones and litter at, again, poor old monitor lizards (in defence, however, not in an attempt to eat them).

The classic tool-user, in that it was one of the first to be observed and recorded, in 1901, is the woodpecker finch of the Galapagos. It plucks a spine off a cactus and, holding it in its beak, uses it to winkle insects out of holes in trees. Using thorns as meat-hooks, shrikes will store their surplus small reptiles, birds and mammals in spiky bushes, and the grey flycatcher of the Serengeti will pursue termites with a blade of grass.

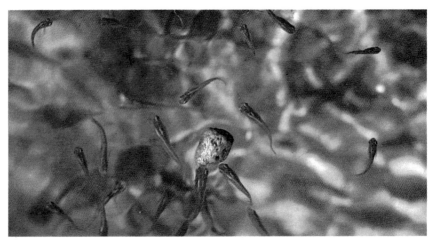

Possibly the most ingenious bird of all is this North American green heron (top). It has learnt to use bread, which people put out for it and which it doesn't particularly like, as bait for fish (centre), which it does like. Sometimes, if no fish come to the bait, the heron will move the bread to another spot (bottom). If there is no bread available, it will even use a feather as a lure. This is not an isolated instance: green herons have been observed doing this in Africa and Asia as well.

If the green heron is an angler, the North American anhinga (top) is a spear-fisherman. Swimming low in the water, with its body below the surface and its neck sticking out like a periscope, it stabs fish with its sword-like beak. The hapless fish is then tossed in the air and caught in its mouth. The classic avian tool-user is the Galapagos woodpecker finch (below), which uses a cactus spine to winkle insects out of holes. If it can't find a spine it will snip the sidepieces off a twig and use that.

The green jay of the southern United States does the same with a twig, and the American brown-headed nuthatch uses one bark-scale to prise others loose and to expose the invertebrates underneath. But the woodpecker finch still goes one step farther than any of these: if it can't find a cactus spine and has to make do with a twig, it will streamline it by snipping off side-branches. This makes it not only a tool-user but a tool-*maker*.

Tool-making suggests a degree of foresight that birds might not normally be credited with, but consider this: in Florida there is a green heron which lives in a park where people occasionally feed the birds, and, not having a great taste for bread and suchlike, it has learned – by a process that scientists term 'insight learning', or figuring things out – to float a piece of bread on the surface of the water and to catch the fish that it attracts. The same trick has been observed in Africa, not only as practised by green herons but also by pied kingfishers. The Florida heron, though, has now been reported to have taken the step from real bait to the artificial lure: it picks up a feather and drops that on the water, and this works just as well.

But the best of the avian tools have been made by evolution. These are the very specialised feeding implements known as beaks. In most cases it is possible to judge by the beak's shape what a bird mainly eats. The 'standard' bill, a basic pair of forceps, is used by the generalised feeders for picking up any of a wide range of food items. This is what the blackbird and the other thrushes have, as well as the gulls, crows and starlings: opportunistic birds, adaptable birds, those that have fared best in a drastically changing environment.

The thin, tweezer-like, surgical-looking bills indicate insect-catchers. These are often species that have to capture their prey on the wing, such as flycatchers and swallows. Birds that stalk after fish – classically, the herons (including that feathered Sherlock Holmes, the black heron, which, to make shade as it tracks through the shallows, fans its wings out and forward over its head, looking for all the world as if it were drawing up a cloak) – have long, spear-shaped beaks. Fish are taken at the very end of the bill, in a lightning-like snap. One of these birds, the cormorant-like anhinga of the tropical and subtropical Americas, Africa, India and Australia, actually spears its fish, which it then tosses into the air and catches. Some ducks, such as the red-breasted merganser, are known as 'sawbills' because the inner edges of their mandibles are serrated, the better to grasp fish with. But most other ducks don't have serrations and use their bills to sieve the water for tiny particles of vegetable matter. The spoonbill, which is slightly misnamed, since its mandibles are not concave like a spoon but flat like a wooden ice-cream digger, holds its beak slightly open and, as it strides steadily through the water, sweeps it from side to side and snaps it closed on any insects, molluscs, small fish or other creatures that make themselves felt. The avocet uses its distinctive upturned bill in much the same way, scything the water as it paces forward.

Some of the largest beaks in relation to the size of their owners belong to waders, such as the snipe and the woodcock, and are used to probe for small organisms in the sand and mud. The oystercatcher has a long, straight bill, which is much thicker than most because it is needed to chisel open mussels. Turnstones have a crowbar of a bill, for upending pebbles, as does the wrybill plover of the southern hemisphere, for probing under stones, except that it is curved to the right.

Owls, hawks, falcons, eagles and especially vultures have a set of butcher's instruments, and woodpeckers have a combination of a hammer and a pick — carpenter's tools. The finches, of course, have nut-crackers, but the most vice-like of these is the European hawfinch's, which can split open an olive stone, a feat that requires more than 45 kilos of pressure. Hummingbirds have as many different drinking probes as they have species. And once upon a time in New Zealand there was a bird called the huia, a member of the wattlebird family, which had a different bill shape for either sex. The male had a woodpecker's drill, and the female had a long, thin hook. One probed, the other dug, and both shared. And, like so many island birds that have received a visitation from man and his commensals, both are now extinct.

Tongues also have special adaptations. Anyone who has ever had a pet parrot will have noticed how it uses its tongue much like a thumb when it picks up a nut. Another well-known tongue is the woodpecker's, which, as soon as the bark is chiselled away and the insect boreholes are exposed, is inserted as a flexible probe. Though most birds' dominant senses are sight and hearing, taste of course counts too. Many insects defend themselves by tasting bad, and advertising their noxiousness with bright colours. This means that once a bird has naively picked up, say, the yellow and black caterpillar of the cinnabar moth and then spat it out in disgust, it will remember the bad experience and thereafter

Among the most extreme of the specialised beaks are those of the Australasian helmeted honey-eater (right), which sips nectar; the South American toco toucan (below), which uses its beak's extraordinary length to reach fruit other birds cannot reach; and the Australian palm cockatoos (opposite), whose cracking power no nut is hard enough to resist.

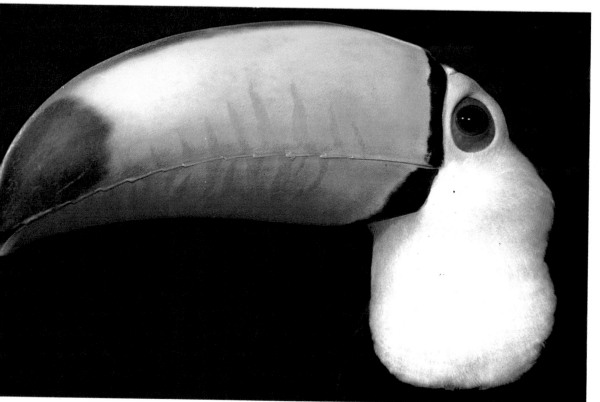

leave anything alone that looks as if it's wearing a football jersey. The same goes for most insects which sport bright colours, as the alternative to camouflage, although some perfectly palatable insects have picked up the trick of looking like the nasties – an evolutionary strategy that puts sheep in wolves' clothing.

Ordinarily, though, birds use taste much as any other animal: in the first instance as a guide to the young on what is and isn't fit to eat, and thereafter as an indicator of poisons and of any food's nutritional component. Touch is also used by birds that have to poke around in dark places – under mud, in murky ponds – and the sense of smell is developed in a few species, almost as a last resort. Oilbirds and kiwis can smell, as can storm petrels and other of the so-called tubenoses, which appear to use odour to identify their mates and their home burrows; the individuality of these odours is so distinctive that even the ornithologists who are studying them can be guided by it. American turkey vultures, which are not related to the other birds of prey, as the Old World vultures are, but to storks, can smell – a talent that helps them find carrion lying beneath the canopy of thick forests. Often the other, non-smelling American vultures have no choice but to wait in the treetops until the turkey vultures have discovered where dinner is.

This is also one example of a mild degree of the parasitism which runs throughout feeding strategies in the whole class of birds. Most birds, in fact, do rely on their own wits and senses to feed themselves, but a large minority have devised various ways of cheating. The pirates of the bird world are the skuas of the polar regions and the frigatebirds of the tropics. The great skua of the North Atlantic has its fishing done by terns, gulls and gannets. The parasite waits at the colony for one of the seabirds, which may have travelled as far as 500 kilometres, to appear with a load of fish, and when it does the skua soars after it, chases it and, finally, after a great display of aerobatics, catches the end of its wing and tips it towards the sea. To make the skua let go, all the other bird can do is drop its catch, which the skua seizes just before the fish hits the water. The frigatebird, one of the fastest-flying and most adroit of the tropical seabirds, does the same thing to boobies, pelicans, cormorants, gulls and even the speedy tropicbirds – always, like the skua, swooping down to catch its meal before it is lost in the sea. Thus do bald eagles rob ospreys, and even blackbirds rob song thrushes of their de-shelled snails.

But there is a more benign way of taking advantage of other creatures. It is not quite symbiosis, in that the one species does not necessarily do anything *for* the other, but it doesn't hurt it either. This, again, is commensalism. It is the relationship of the North American cowbird (formerly buffalo bird, until the West was won) to the cow, which it uses as a beater, walking alongside it and catching the insects that it stirs up. The ubiquitous cattle egret does the same, in both hemispheres, accompanying elephants in Africa, water buffalo in southern Asia and cows in North America.

It happens that some cattle egrets do not have this habit, and according to one South African study, the egrets which shun the practice receive a third less food per hour. But the cattle egret has to expend energy walking. The carmine bee-eater doesn't even do that; it rides on its beater's back. Its favourite 'host' is the large, near-flightless kori bustard, but it is also seen on camels, antelopes and ostriches. An extension of this habit into full-fledged symbiosis is the oxpecker's bargain with several of the large African ungulates: it rids them of ticks in exchange for the blood the ticks contain.

There is one instance of symbiosis between a bird and humans that, as far as anyone knows, was instigated by the bird some time in the distant past and involves a strategy that is specific to people, not as house-builders or rubbish-producers, but as fellow animals with a common interest. The African black-throated honey-guide is unique in the animal kingdom in that it can metabolise beeswax. But to extract it is a long and difficult process with a reward which barely covers the bird's energy loss. Somehow, some time — and this would almost be more believable as a Kipling 'just-so' story — it observed that beeswax was a lot easier to exploit if people first opened the hive to take the honey. So the bird began to search out hives and then to fly to the nearest human settlement and tell the people what it had found, and it told them with a call that imitated the buzzing of bees. The people would follow the bird to the hive, take the honey and let the bird have the beeswax. And it still happens. What this means, presumably, is that there is at least one creature on the planet for which the disappearance of humanity would be almost as disastrous as the disappearance of the dodo was for the tambalacoque tree.

Bird parasitism ranges from the almost friendly to the downright rude. Cattle egrets (above) are no trouble at all to the large mammals whose feet act as insect-beaters for the birds, nor is the carmine bee-eater (above right) a pest to the kori bustard, which it uses for the same purpose, as well as for transport. But when a sheathbill (right) takes food from the mouth of a baby adélie penguin, despite the admonitions of the parent, it is being impolite, to say the least.

Birds which do not know the meaning of stealth include the blue and yellow macaws (left). Like all parrots, they are principally fruit-eaters, and don't have to worry about their prey hearing them. As a result, they are among the noisiest and most colourful birds on earth. The Malaysian fish owl (below), unusually for an owl, has no need for stealth because its prey is all under water and unable to hear its loudly flapping wings.

'Sea-ward, white gleaming through the busy scud
With arching wings, the sea-mew o'er my head
Posts on, as bent on speed: now passaging,
Edges the stiffer breeze; now yielding, drifts;
Now floats upon the air, and sends from far
A wildly-wailing note.'

Samuel Taylor Coleridge

Chapter Six
HOME ON THE WIND

When birds discovered migration by flight, they came nearer than any creature ever had to correcting the anomaly of the earth's $66\frac{1}{2}$-degree tilt. This accident of nature could go on shifting the good weather every year from one end of the planet to the other, but here were creatures with a mobility that would never be surpassed. If the sun went south for the winter, so — with comparatively little effort — would they. It was as though the animals of the air became part of the very atmosphere, shifting with the wind as the seasons changed.

The advantage of migration is that the bird can take itself to the best possible spot for its biological needs. A barnacle goose, which winters in the northern cool-temperate zone, in Britain for example, can choose to reproduce in the short summer of the high Arctic, with an abundance of protein in the form of insects for the newly hatched goslings to grow up on. The geese can enjoy days that are so long as to obliterate night altogether, as well as relative safety from both ground and aerial predators.

A swift, whose northern spring migration in many cases stops south of the barnacle goose's wintering grounds, comes north, like the goose, for longer breeding days and for a seasonal abundance of aerial insects of which the local resident birds are unequipped to take advantage. In the winter, the swift goes with the sun and the food to central and southern Africa.

Both birds migrate, but one is essentially a cold-weather species and the other a warm-weather one — and each maintains its preferred weather by keeping itself at more or less the same angle to the sun. One bird which has evolved to take advantage of as much daylight as possible uses migration to ensure that it is in the sunshine more than any other creature on earth. This is the Arctic tern, and it achieves this distinction by undertaking the longest of all migrations. Sometimes called 'swallows of the sea', the terns breed near the Arctic Circle, and when the

Both the snowy owl (opposite) and the Arctic tern (left and below) nest in the far north, but while the owl only moves a short way south in the winter, the tern flies all the way to the Antarctic — the longest migration of any bird.

days start to shorten they fly to the shores of Antarctica, some 13,000 kilometres away – if you follow the route of the proverbial crow. If you follow the route of the actual Arctic terns, like ancient sailors they seem to want to keep a coastline in sight as much as possible and otherwise to detour into the path of prevailing tailwinds, and so the round trip can easily amount to 35,000 kilometres.

Terns nest along all the Arctic coasts, and whether they head for the southern seas by way of the Atlantic or the Pacific depends, of course, on which of the oceans they are nearer to when they start. The Pacific route follows the contours of the west coast of the Americas, and the Atlantic route follows the European west coasts and the west coast of northern Africa. A bird that has come this far has been travelling for about a month, day and night, at a speed of approximately six kilometres per hour, plus help from the wind, and when it reaches Cape Horn or the Cape of Good Hope, it is close to exhaustion. But here it encounters some of the strongest and most reliable winds of the entire trip. The continual easterly current of air which escorts a similar current of water – together known as the Roaring Forties – around the world at that belt of sea uninterrupted by any land of significance, enables the terns to rise high on the winds. They surrender to the air flows, allowing themselves to be carried towards the pack ice.

Included among the arrivals are, of course, the young birds, the ones which had just finished fledging in the Arctic and were using their wings and feathers for the first time when they took off on the epic journey. They are the most tired and least likely to impose any intention of their own on the whims of the Forties. As a result, it is often they that get carried the farthest, and a young bird which had come down the Atlantic route might find itself finally ashore somewhere south of the Pacific. So, when it flies north again in the southern autumn and northern spring, for all practical purposes it will have circumnavigated the globe, in its first year of life.

As prodigious a feat of flight as the Arctic tern's may be, for total adaptation to the planet's gaseous medium even this marathon champion has nothing on the swifts. Whatever else it can do, the Arctic tern can still get tired from time to time, and it may have to rest on the sea for a while. But a swift, except when it is on its nest, is always in the air – and many of the migratory swifts only have nests for a few months of the year, and only sit on them when they are laying or incubating eggs.

Swifts are on the wing when they catch and eat their prey — flying insects, floating spiderlings, any of the so-called aerial plankton — when they are gathering nesting materials, even, often, when they are copulating. They can fly and sleep at the same time, the wings working while the rest of the bird is dropping in and out of a series of catnaps. When a swift takes a bath, it simply flies into the water and then re-emerges, dripping but still flapping.

Almost all of the eighty-two species of swifts are tropical birds, and even that familiar northern European symbol of summer, the common swift, is really most at home over the African savannahs and rainforests, where it spends the bulk of the year. The swift that comes to Europe is, compared with other swifts, a bird that nests in the far north, and a long-distance migrator. Indeed, early and late in the season, the northern weather can occasionally be a problem, particularly when cold depressions blow up. But if anything can fly, a swift can, and when unseasonal cold weather arrives, the bird just gets out of the way, even if it means flying hundreds of kilometres.

The strategy of nesting in the northern summer has proved very successful, and the common swift is one of the most numerous of all its family. In Europe it finds a wealth of seasonal insects, miraculously long days in which to indulge in aerial grazing, and a limitless choice of the cave-like crannies where it likes to nest, almost invariably provided by humans nowadays. Gravity-defiers that they are, when swifts approach *terra firma* it is from the side. On the rare occasions when they feel the need to alight, they grasp the side of a vertical surface: a swift cannot stand on the ground. These northern footholds are usually man's buildings, of course, and for that reason we tend to associate swifts with two other harbingers of summer, house-nesting birds which also always return to the same places, which are also agile fliers and which even have the swifts' streamlined shape. These, of course, are the swallows and the martins. But, oddly, the swifts are not related to them. Instead, they are classed with those other small, specialist, almost insect-like acrobats, the hummingbirds.

And like hummingbirds, which seem blindingly quick even when they are going nowhere at all, swifts have a very deceptive flight speed. They are good at manoeuvring and they are hard to watch, but they are, in fact, not particularly swift. Their top speed when flitting about after insects is 23 kilometres per hour, which makes them one of nature's

Swifts are the most aerial of
warm-blooded creatures,
doing everything on the
wing except nesting.
Between the time a young
swift launches itself and is
old enough to nest and thus
needs to land again, two or
three years can pass.

slower birds, in about the same class as a blue tit. Such an undisting-uished flier as a crow can do 50 kilometres per hour, a house sparrow can go up to 40, as can most gulls. And the fastest bird of all, under any circumstances, is a peregrine falcon in a stoop: one has been timed at 180 kilometres per hour.

But so what? Swifts may not fly fast, but flying is all they do, and when they are migrating high on the air currents, asleep and awake, they can manage a steady 40 kilometres per hour. And these currents tend to be in the same places, at the same strength, in the same direction and at the same time every season – and the birds that use the winds have evolved to be there as well, and are as predictable as the currents.

Some birds have migration routes as fixed as human roads. But birds of different species use the winds in different ways, depending on such factors as the shapes of their wings. They also have different reasons for migrating and, indeed, are coming from and going to different places. So the routes can be like dirt tracks used by one or two species at the beginning or end of a journey; or like secondary roads carrying, say, only thrushes or oscine insectivores or birds of prey; or like motorways of wonderfully consistent and uplifting currents that also happen to run, for example, due south from Europe to Africa – exactly where most birds are going. These latter will become thick with birds of all species and will provide, among other things, some of the greatest birdwatching spectacles on earth.

Two of these places happen to be at either end of the Mediterranean: the Bosphorus and Gibraltar. In both cases, the birds belong to a wide range of species that breed in Europe and winter in Africa, and to be in either place from September to November is to witness a constant overhead procession of whitethroats, blackcaps, warblers, martins, swallows, swifts, common cranes, white storks, common buzzards, honey buzzards, black kites, ospreys and falcons. These birds depend upon thermals and save much energy by riding on them. Most other birds, provided the weather is good, perform a broad front migration straight over the Mediterranean.

North America, which lacks such a landlocked sea with all its implications for weather and wind, not to mention absence of resting places (for few birds have the stamina of an Arctic tern), can be divided into four principal 'flyways'. An assortment of small passerines and a few other birds enter the flyways almost indiscriminately, so that any species of these can be found in any of the four. The Pacific flyway runs just west of the Rockies, the Central flyway just east of them. The Mississippi flyway covers the Great Plains and the Atlantic flyway the eastern seaboard, and all four contain more or less the same large collection of waterfowl – millions of waders, geese and ducks. Many of the smaller, land migrants can feed almost anywhere, and have no need to fly from one wetland 'hotel' to the next.

In the Pacific flyway most of these are heading for the lakes and shores of Southern California, for the Sea of Cortez or for middle Mexico and beyond, but birds migrating down the other three mainly come to roost on the rich shores of the Gulf of Mexico, which, because of its irregular, reedy coast, its warm climate and its multitude of barrier islands, has a concentration of wintering waterfowl unequalled almost anywhere. Some of the warblers and ospreys, and all of the martins, swifts, hawks, flycatchers, tanagers, vireos and thrushes, go on by way either of Mexico or the Antilles to the Central and South American rainforests, but almost all of the waterfowl stay to graze and fish in the Gulf's lagoons, estuaries and reedbeds, or on nearby stubble fields.

Perhaps most famous of these is the whooping crane, the object of one of the most intensive conservation campaigns ever mounted. The total now numbers just over 100 in two flocks, which is very high compared with the sixteen it had dropped to when the species' last nesting site was discovered in 1955.

The recovery is also a tribute to the bird's luck in having survived America's age of rampant slaughter just long enough to reach the first glimmerings of a new age of reluctant apology. In former years at least three species of extremely plentiful birds – the passenger pigeon, the Carolina parakeet and the Labrador duck – were wiped out. The heath hen defied all efforts to save it and died in the arms of latter-day conservationists, and several other birds have, like the American bison on the ground, been reduced to museum-like populations kept going only by conscious human effort. Among the very rare are the ivory-billed woodpecker, which has not been seen on the North American mainland since the late 'sixties (though there was a reported sighting in Cuba in May, 1986); the California condor, the continent's largest bird and one that had the misfortune to have a home range which came to include the conurbation of Los Angeles (a factor that has helped reduce its population to less than twenty, all but five now in zoos); the Eskimo

Carolina parakeets (opposite) and passenger pigeons (left) have to be depicted by paintings because the last of each species died in 1914. Both were once plentiful, but both fell victim to the North American pioneer spirit, which viewed wildlife as either an inexhaustible resource or a nuisance. The pigeon was hunted as food, the parakeet as a pest.

curlew which, like the passenger pigeon, used to migrate in sun-obliterating flocks but also had (and in its tiny relict population still has, if any birds really do exist) one of the most peripatetic itineraries on earth – Alaska and western Canada, to eastern Canada and the United States, to the Caribbean, to the Rio Plata in South America and then back to Alaska via Mexico, Texas and the Midwest – exposing itself on the way to every shotgun on two continents; and two kinds of prairie chicken, Attwater's and the greater, most of whose necessary prairies are now cornfields.

It should not be forgotten that the people we now call Americans are all within a few generations of having come from somewhere else in the world, mainly Europe, and that the hemisphere was 'civilised' – turned into an imitation Europe – in a virtual blink of the eye. It never occurred to any of the new arrivals that there might be any other way of dealing with the bonanza. Most of the birds and other animals that were forced into literal or practical extinction were the victims of agriculture which took over their habitats, of 'pest'-control programmes or of a degree of overhunting that transcended greed until it resembled something like a blood frenzy. This last is what, in particular, befell the passenger pigeon and the Eskimo curlew, which were shot at almost every time they were sighted, with individual hunters boasting of bags in their hundreds or thousands, day after day, during migration time.

In 1878 in Petoskey, Michigan, at one of the passenger pigeon's last large nestings, one man in one season killed $60,000 worth of birds, which at the going price would have been about three million of them. The last passenger pigeon, a female named Martha, died in Cincinnati Zoo in 1914. This was once probably the most plentiful bird in North America, and to have wiped it out with the gun is one of mankind's greatest feats of concentrated destruction.

The ivory-billed woodpecker, the prairie chickens and the condor have simply gone the ways of the rest of their ecosystems. The Carolina parakeet lost its habitat – the forests of the south-eastern United States – to farms, but it might have survived had not the farmers, as farmers will, objected to the birds eating the products of the fruit trees growing where the parakeets' own fruit trees had once grown, and had not the parakeets, for some reason, evolved the habit of flocking around injured or dead comrades. This made them a farmer's dream of a pest, because a whole orchard-full could be killed almost at once.

Why an animal would concentrate on the scene of an accident rather than getting away as fast as possible is the subject of some discussion among animal behaviourists, but one theory is that in the world in which the birds evolved, a world that did not contain armed farmers, a large flock could confuse a predator and drive it away. There are also some arguments about the genetic value of siblings and cousins and about the similarity of behaviour in whales and dolphins – which will try to hold an injured companion upright, a boon to the busy whaler. But, for whatever reason the parakeet behaved in this way, the very fact that it existed in good numbers before the arrival of civilisation meant that it had been a perfectly good strategy for millions of years. Anyway, it was also hunted for meat, was collected in large numbers for the pet trade (though not bred, apparently) and was forced out of its nesting holes by another new arrival, the European honeybee. The last reliable sighting of a wild Carolina parakeet was at Lake Okeechobee, Florida, in 1904, and the last captive bird died in the same year as the passenger pigeon and at the same zoo: Cincinnati, 1914.

The whooping crane, too, suffered for all of the same reasons. It had once nested in a broad swathe of wetlands stretching from what is now the Canadian Northwest Territories across Alberta and Saskatchewan and through Minnesota to Iowa and Illinois, and it wintered along the entire northern Gulf coast. But, as soon as the Europeans reached the prairies, they drained all the standing water and planted grain crops (the whole region is now known as the world's breadbasket, and its surplus produce fills millions of square metres of storage space). That took care of the habitat. Otherwise, they were shot as they migrated.

After 1922, although they were occasionally sighted (and still sometimes killed), many years passed without anyone finding a whooping crane nest. In 1939, some were seen in Louisiana, which meant that a few birds were so desperate that they were trying to breed in their wintering grounds, but those nests disappeared the next year. Then, in 1955, in the Northwest Territories' Wood Buffalo National Park, at the extreme north end of the crane's original range, the last fourteen birds were found, with nests, trying to breed. They were then traced to their wintering grounds, Texas's Aransas National Wildlife Refuge, and a new era of American wildlife conservation began.

The birds became a national issue in both of the countries involved. They are strictly protected at either end of their migration, and so much

The whooping crane (right), once nearly extinct, has become a symbol of successful conservation. It migrates between Canada and south-eastern Texas. The red-tailed tropicbird (below) doesn't migrate, but it flies prodigious distances over the oceans in search of food.

The frigatebird (left) can fly and display at the same time. It can also rob tropicbirds returning from their great oceanic expeditions — a manoeuvre that requires much agility. The fulmar (below) spends a lot of time at sea, too.

publicity has been given to them and their plight that no hunter could ever mistake one for something else, although at least one has been shot in this decade. To get some of the cranes' eggs, as it were, out of the one basket, the Canadian and United States wildlife authorities began a programme of fostering whooper hatchlings with flocks of the other American crane – the sandhill – so that the youngsters would learn new breeding and summering grounds and new migration routes – this one between northern Alberta and New Mexico. It worked, and now there are 125 cranes in both flocks, not to mention whooping crane societies and newsletters and a whole conservation movement based on the moral and practical lessons of the experience, from which many animals and plants have benefited. (In 1979 in Houston, Texas, a birdwatcher was charged with attempting to kill his son-in-law with a dining-room chair, after the son-in-law had jokingly told him that the Christmas goose he had just eaten was a whooping crane.)

At least one Asian crane has benefited from the whooping crane experience – the Siberian crane – which is only slightly less rare and which migrates, as do many Asian species, between northern Siberia and both India and southern China. These cranes are in more danger as they fly on migration, when they may be hunted, and in at least one of their resting places, which is part of a very contentious piece of territory in the Afghan war, than in their remote breeding area. Taking advantage of this, the Soviet government used officials and advice from the United States Fish and Wildlife Service to establish another flock in western Russia.

Asian migrants generally move more or less due south from their breeding areas in Siberia, China, Japan and the Soviet Far East, wintering in East and southern Africa, southern Arabia, the Indian subcontinent, South-east Asia, Indonesia, Papua New Guinea and sometimes even Australia.

As with feeding strategies, there are as many ways of migrating as there are migrants. The distances involved can be as long as an Arctic tern's or as short as that of some of the song thrushes that breed in England and winter in Ireland, or the shelducks that travel between the Scottish islands and the coast of Germany. Even Canada geese, which in their native land are prodigious migrators, have after introduction to Britain restricted their migration to the area between Land's End and John O'Groats. Puffins and some tropical island birds migrate from east to west, with food abundance as their motivation, rather than climate.

And many birds don't migrate. A number of biologists argue that a so-called 'sedentary' bird is one better adapted to its habitat and to life in general, than one that is always on the move. But this argument fails to take into account that movement is in itself not time out of time but is instead living in the air, most birds' home medium. A swift that is not on the wing, unless it is attending to its necessarily stationary nest, is probably ill, and there is no question that a bird that flies so high and fast and effortlessly is not going to fly to the place where it can feed best and then fly back to the place where it can nest best. Another argument proposes that migration habits are leftovers from times when the continents were closer together, and the birds simply extended their journeys imperceptibly as the continents imperceptibly drifted. While this is probably a good explanation for the habits of certain other migratory animals— marine turtles especially— examples such as that of the British version of the Canada goose and certain other introduced or curtailed species show that birds have more adaptability and logic. Probably the best way of looking at it is the common-sense one: everything adapts as best it can. An organism, in fact, *is* adaptation.

When winter comes, our blackbird's adaptation is to stay put. It knows its territory and it can feed on stored chaff and moss and certain insects. And if the winter is not too hard, and its fat supply holds out, the bird will probably survive to sing in another spring, to steal a march on all the birds still luxuriating in Africa or the South of France— and to produce in the year as a whole an extra clutch of two or three offspring. For the species needs the extra production: the winters are *not* always easy, and very many do die. There are some birds, of course, which, far from dreading winter, seem to consider it their natural season. The snowy owl springs to mind, but in fact it still likes to travel slightly south out of the worst of the Arctic weather. Probably the most naturally wintery bird is the ptarmigan, which changes its camouflage in winter to snow white, as does the willow grouse of northern Scandinavia, except for its head and neck. This bird not only burrows under the snow, but has developed a technique of lowering its body temperature to as near zero as any warm-blooded creature can. The evolution that created self-heating creatures able to live in cool climates produced at the other end of the spectrum one that could make itself cold as a *defence* against the cold— the opposite and equal of fighting fire with fire.

One of the best-adapted birds to winter is the ptarmigan (above). Its plumage changes to white, and it can sit out blizzards by burrowing under the snow. Many non-migrating birds, such as the cock bullfinch (right), will not survive until spring if the winter is unusually hard.

The cold of winter affects birds less than a shortage of food. When a morsel is found, it can lead to squabbles, such as the one among the grey herons (left), one of which has found a fish. As the season drags on, birds such as the rooks (below) simply save energy by sitting still.

Acknowledgements

I have to thank a remarkable editor called David Helton for knocking into shape the words I wrote for this book and the shooting scripts I wrote for the *Birds for all Seasons* series (plus some other material from earlier programmes on television). David is a Texan who wrote a novel about a guitarist with six fingers on each hand, was assistant editor of the conservation journal *Oryx*, and now works for the monthly magazine *BBC Wildlife*. More relevantly, he is a skilled and hard-working wordsmith and a responsible and sensitive science writer. The felicitous English, readability and shape of the text is his doing. I value him as both colleague and friend.

The responsibility for what now appears is of course 100 per cent mine. At an earlier stage Mrs Anne Norris edited a sample chapter and Wendy Dickson typed my 60,000 words – more than once! My family kindly decamped to Sussex one Christmas leaving me alone, immersed and unshaven, for eleven productive days.

Ros Kidman Cox, the editor of *BBC Wildlife* and another good friend, advised on additional sources of photographs. Jennifer Fry assiduously followed up my initial appeal to each photographer, putting every possible effort into getting the best. David Helton and I both worked very closely with Hilary Duguid, our editor at BBC Publications, and with Ann Thompson, who is responsible for the volume's beautiful design. Appropriately, the index was compiled by Richard – Dickie – Bird, and the proofs were checked with great thoroughness by Bill Games. Overseeing all has been Tony Kingsford, BBC Books Editor.

In what I recognise as a break with tradition I am not including in these acknowledgements *all* those who helped with the programmes rather than directly with the book. As they know, they have already been thanked, and in what I hope they consider appropriate ways.

Those most directly involved in the production of the television films are: Jeffery Boswall (Series Writer and Executive Producer); Adrian Warren (Producer); Pru Palmes and Mike Catsis (Research); Kaija Pepper and Diana Williams (Production Assistants); Magnus Magnusson (Narrator); Doug Allen, Andrew Anderson, Eric Ashby, Hans Bang, Stephen Bolwell, Rodney Borland, Rob Brown, Gary Capo, Theo Cockerell, Red Denner, Robert G. Dickson, Ron Eastman, Kevin Flay, Chris Fryman, Len Gilday, Mike Herd, Jeremy Humphries, Rodger Jackman, Alan Kemp, Robert Long, Hugh Maynard, Hugh Miles, Gunnar Nilsen, Mike Potts, Richard Price, Neil Rettig, Michael Richards, Chris Strewe (Wildlife Cameramen); Chris Orrell and Peter Heeley (Film Editors); Derek Whelan and Teresa Hughes (Assistant Film Editors) and Linda Thomson (Opening Titles Designer).

Reading and Listening

This appendix comprises a bibliography and a discography, with entries listed region by region.

We are all familiar with the way geographers divide the world into continents: by land mass. Ornithologists (along with other zoologists) divide the world according to animal communities – not always by sea boundaries but a different kind of physical boundary or barrier, for example, a desert (notably the Sahara) or a mountain range (notably the Himalayas). This is clear from the map of the zoogeographical regions of the world below.

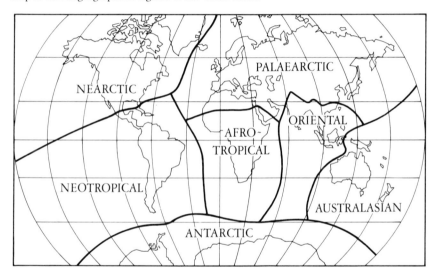

READING

PALAEARCTIC REGION   Europe, North Africa, temperate Asia

A Synopsis of the Birds of China by Cheng Tso Hsin. Science Press, Beijing (Peking), 1986.
Handbook of the Birds of Europe, the Middle East and North Africa: the Birds of the Western Palearctic,
  vols I–V, by S. Cramp and others. Oxford University Press, Oxford, 1978–86.
Birds of the Soviet Union, vols I–VI by G. P. Dementiev and others. Jerusalem, 1966.
The Shell Guide to the Birds of Britain and Ireland by J. Ferguson-Lees, I. Willis and
  J. T. R. Sharrock. Michael Joseph, London, 1983.
A Field Guide to Birds of the USSR by V. E. Flint, R. L. Boehme, Y. V. Kostin and
  A. A. Kuznetzov. Princeton University Press, Princeton, N.J., 1984.
The Birds of Oman by M. Gallagher and M. W. Woodcock. Quartet, London, 1980.
The New Where to Watch Birds (in the UK) by J. Gooders. Deutsch, London, 1986.
Birds of Britain and Europe by N. Hammond and M. Everett. Pan, London, 1980.

A Field Guide to the Nests, Eggs and Nestlings of British and European Birds, with North Africa and the Middle East by C. J. O. Harrison. Collins, London, 1975.

An Atlas of the Birds of the Western Palaearctic by C. J. O. Harrison. Collins, London, 1982.

Birds of Britain and Europe with North Africa and the Middle East by H. Heintzel, R. S. R. Fitter and J. Parslow. Collins, London, 1972.

Popular Handbook of British Birds (revised edn) by P. A. D. Hollom. Witherby, London, 1980.

The Atlas of Wintering Birds in Britain and Ireland by P. Lack. British Trust for Ornithology, Tring, Hertfordshire, 1986.

A Field Guide to the Birds of Britain and Europe (4th edn) by R. T. Peterson, G. Mountfort and P. A. D. Hollom. Collins, London, 1983.

The Birds of China by R. M. de Schauensee. Oxford University Press, Oxford, 1984.

The Atlas of Breeding Birds in Britain and Ireland by J. T. R. Sharrock. British Trust for Ornithology, Tring, Hertfordshire, 1976.

A Field Guide to the Birds of Japan by the Wild Bird Society of Japan. Wild Bird Society of Japan, Tokyo, 1982.

AFRO-TROPICAL REGION   Africa south of the Sahara

The Birds of Africa, vols I and II by L. H. Brown, E. K. Urban, S. Keith and K. Newman. Academic Press, London, 1982.

Roberts' Birds of South Africa (4th edn) by G. R. McLachlan and others. Struik, Cape Town, 1978.

The Birds of Southern Africa by K. Newman. Macmillan, Johannesburg, 1983.

The Birds of the Seychelles by M. Penny. Collins, London, 1974.

A Field Guide to the Birds of West Africa by W. Serle, G. J. Morel and W. Hartwig. Collins, London, 1977.

A Field Guide to the Birds of East Africa by J. G. Williams and N. Arlott. Collins, London, 1980.

ORIENTAL REGION   Tropical Asia

Field Guide to the Birds of the Eastern Himalayas by S. Ali. Oxford University Press, Delhi, 1977.

A Pictorial Guide to the Birds of the Indian Subcontinent by S. Ali and S. D. Ripley. Bombay Natural History Society/Oxford University Press, Delhi, 1983.

Handbook of the Birds of India and Pakistan by S. Ali and S. D. Ripley. Oxford University Press, Delhi, 1984.

The Birds of the Malay Peninsula, Singapore and Penang by A. G. Glenister. Oxford University Press, Kuala Lumpur, 1971.

A Guide to the Birds of Nepal by C. and T. Inskipp. Croom Helm, London, 1985.

A Field Guide to the Birds of South-East Asia by B. King, M. Woodcock and E. C. Dickinson. Collins, London, 1975.

AUSTRALASIAN REGION   New Guinea, Australia, New Zealand and Pacific Islands

Birds of New Guinea by B. M. Beehler and others. Princeton University Press, Princeton, N.J., 1986.

The New Guide to the Birds of New Zealand by R. A. Falla, R. B. Sibson and E. G. Turbott. Collins, Auckland and London, 1979.

A Field Guide to the Birds of Australia by G. Pizzey. Collins, Sydney, 1980.

The Birds of Australia: a book of identification by K. Simpson and N. Day. Croom Helm, London, 1985.

A Field Guide to Australian Birds by P. Slater. Vol I, Oliver and Boyd, Edinburgh, 1971; vol II, Scottish Academic Press, Edinburgh, 1975.

NEARCTIC REGION   North America south only to northern Mexico

The Audubon Society Master Guide to Birding (3 vols) by J. J. Farrand. Knopf, New York, 1983.

The Birds of Canada by W. E. Godfrey. National Museum of Canada, Ottawa, 1966.

Field Guide to the Birds of North America by the National Geographic Society. NGS, Washington, 1983.

A Field Guide to the Birds East of the Rockies (4th edn) by R. T. Peterson. Houghton Mifflin. Boston, Mass., 1980.

Birds of North America (expanded revised edn) by C. S. Robbins, B. Bruun and H. S. Zim. Golden Press, New York, 1983.

NEOTROPICAL REGION   Central and South America

Birds of the West Indies: a Guide to the Species of Birds that Inhabit the Greater Antilles, Lesser Antilles and Bahama Islands by J. Bond. Collins, London, 1979.

A Field Guide to the Birds of Galapagos (revised edn) by M. Harris. Collins, London, 1982.

A Guide to the Birds of Colombia by Steven Hilty and William Brown. Princeton University Press, Princeton, N.J., 1986.

A Field Guide to Mexican Birds by R. T. Peterson and E. Chalif. Houghton-Mifflin, Boston, 1973.

A Guide to the Birds of Panama by R. Ridgely. Princeton University Press, Princeton, N.J., 1976.

A Guide to the Birds of South America by R. M. de Schauensee. Academy of Natural Sciences of Philadelphia, 1982.

A Guide to the Birds of Venezuela by R. M. de Schauensee and W. H. Phelps. Princeton University Press, Princeton, N.J., 1978.

ANTARCTIC REGION   The area south of the Antarctic convergence (roughly 50° south)

Birds of the Antarctic and Sub-Antarctic by G. E. Watson. American Geophysical Union, Washington, D.C., 1975.

GLOBAL AND GENERAL WORKS

A New Dictionary of Birds by B. Campbell and E. Lack. T. A. D. Poyser, Stoke-on-Trent, 1985.

The Dictionary of Birds in Colour by B. Campbell. Joseph, London, 1974.

Birds of the World: a Checklist by J. Clements. Croom Helm, London, 1981.

Avian Biology (vols I–VII) by D. S. Farner, J. R. King and K. C. Parkes. Academic Press, New York and London, 1971–83.

A Checklist of the Birds of the World by E. S. Gruson. Collins, London, 1976.

Vanishing Birds: their Natural History and Conservation by T. Halliday. Sidgwick and Jackson, London, 1978.

Bird Families of the World edited by C. J. O. Harrison. Elsevier-Phaidon, Oxford, 1978.

A Complete Checklist of the Birds of the World by R. Howard and A. Moore. Oxford University Press, Oxford, 1980.

Bird Life: an Introduction to the World of Birds by C. M. Perrins. Elsevier-Phaidon, Oxford, 1976.

Avian Ecology by C. M. Perrins and T. R. Birkhead. Blackie, London, 1983.

The Encyclopedia of Birds by C. M. Perrins and L. A. Middleton. Allen and Unwin, London and Sydney, 1985.

The Birds by R. T. Peterson. Life Nature Library, Time-Life International, Netherlands, 1964.

Ornithology from Aristotle to the Present by E. Stresemann. Harvard University Press, Cambridge, Massachusetts, 1975.

Fundamentals of Ornithology (2nd edn) by J. van Tyne and A. J. Berger. Wiley, New York, 1976.

The Complete Birds of the World by M. Walters. David and Charles, Newton Abbott, Devon, 1980.

LISTENING

PALAEARCTIC REGION   Europe, North Africa, temperate Asia

The Peterson Field Guide to the Bird Songs of Britain and Europe by S. Palmér and J. Boswall. 1969–81. Fifteen 30 cm electronic stereo discs, SR RFLP 5001–15; or sixteen cassettes, SRMK 5021–36. Swedish Radio, S-105 10 Stockholm, Sweden. Available from various retail outlets. 558 species (discs); 612 species (cassettes).

The Bird-Walker by J. C. Roché. 1985. Three cassettes, BW-ANG-3K7-85. L'Oiseau Musicien, la haute borie, St-Martin-de-Castillon, 84750 Viens, France. Available in UK from various retail outlets. 406 species.

British Bird Vocabulary by V. C. Lewis. 1980. Six volumes of two cassettes each. V. C. Lewis, Audio-Visual Aids, Rosehill House, Lyonshall, Nr Kington, Herefordshire. HR5 3HS. 127 species.

A Sound Guide to Waders in Britain by J. Burton and N. Tucker. 1984. 30 cm disc. BBC REC 545. BBC Records and Tapes. Available from retail outlets, not direct from BBC. 33 species.

Birds of the Soviet Union: A Sound Guide, 1–3 Divers and Waders by B. N. Veprintsev. 1982. Three 30 cm discs, Melodiia C(90) 18023/4, 5/6 and 7/8. Collets, Denington Estate, Wellingborough, Northants, NN8 2QT.

Soviet Birds by L. Svensson. 1984. One cassette, LSK81. Lars Svensson, Sturegatan 60, S-114 36 Stockholm, Sweden. 31 species, mainly Siberian.

Stimmen der Vögel – Vögelstimmen Südosteuropas 1 and 2 (Bird Song: The Bird Songs of Southern Europe) by M. Schubert. 1973 and 1984. Two 30 cm discs, Eterna 8.21.611 and 8.22.702. VEB Deutsche Schallplatten, Gross-Berliner Damm 27/31, 1197 Berlin, East Germany. 63 species.

Stimmen der Vögel Zentralasiens I and II (Songs of the Birds of Mongolia) by M. Schubert. 1982. Two 30 cm discs, Eterna 8.22 575–6. VEB Deutsche Schallplatten (as above).

Japanese Nature and Birds by T. Kabaya. 1977. Five 30 cm discs, Columbia GX-7022-26. Nippon Columbia Ltd, Tokyo. 174 species.

160 Wild Bird Songs of Japan by T. Kabaya. 1985. Four cassettes. Yamato-Keikokusha Co. Ltd, 1-1-33 Shibadaimon, Minato-ku, Tokyo 105.

AFRO-TROPICAL REGION   Africa south of the Sahara

Des Oiseaux de l'Ouest Africain (Birds of West Africa) by C. Chappuis. 1974–85. Eleven 30 cm discs. Sound supplements to *Alauda* journal. Disc 1 with vol. 42 (2), 2 and 3 with vol. 42 (4), 4–6 and 9 with vol. 43 (4), 8 and 9 with vol. 46 (4), 10 with vol. 47 (3), 12 with vol. 49 (1) and 13 with vol. 53 (2). (Discs 7 and 11 are on different subjects.) Société d'Études Ornithologiques, École Normale Supérieure, 46 rue d'Ulm, 75230 Paris Cedex 05. Nearly 500 species.

Birds and Lemurs of Madagascar by P. Huguet and J. Roché. 1986. Two cassettes, OM-MAD-2K7-86. Co-production by Audio Patrimoine International and L'Oiseau Musicien. Published by Sitelle. Address as L'Oiseau Musicien (see above). 54 bird species.

Southern African Bird Calls by L. W. Gillard. 1985. Three cassettes. Gillard Bird Cassettes, P.O. Box 72059, Parkview, 2122 Johannesburg. Over 500 species.

ORIENTAL REGION   Tropical Asia

A Field Guide to the Bird Songs of South-East Asia by T. C. White. 1984. Set of two cassettes, NSA C1 and 2. British Library National Sound Archive, 29 Exhibition Road, London SW7 2AS. 138 species.

AUSTRALASIAN REGION   New Guinea, Australia, New Zealand and Pacific Islands

A Field Guide to Australian Bird Song, Parts 1 and 2 by R. Buckingham and L. Jackson. 1983 and 1985. Two cassettes. Bird Observers Club, Box 185, P.O. Nunawading, Victoria 3131. About 140 species.

Papua New Guinea Bird Calls: Passerines and Non-passerines by H. and A. Crouch. 1983. Two cassettes. Papua New Guinea Bird Society, P.O. Box 1598, Boroko, Papua New Guinea. 152 species.

Birds of New Zealand by J. L. Kendrick. 1980. 30 cm disc, Viking VP445. Viking Overseas Ltd., P.O. Box 1431, Wellington, New Zealand. Also available from World Wildlife Fund-NZ, P.O. Box 12200, Wellington North, New Zealand. 37 species.

NEARCTIC REGION   North America south only to northern Mexico

A Field Guide to the Bird Songs of Eastern and Central North America, Second Edition by J. Gulledge. 1983. Two 30 cm discs or two cassettes. Laboratory of Ornithology, Cornell University, 159 Sapsucker Woods Road, Ithaca, New York 14850. 245 species.

A Field Guide to Western Bird Songs by P. P. Kellogg. 1962. Three 30 cm discs or two cassettes. Laboratory of Ornithology (as above). 515 species.

Guide to Bird Sounds by J. L. Gulledge and W. W. H. Gunn. 1985. Two cassettes. Laboratory of Ornithology, Cornell (as above). 176 species.

NEOTROPICAL REGION   Central and South America

Voices of Neotropical Birds by J. W. Hardy. 1975. 30 cm disc, ARA-1. ARA Records, J. W. and C. K. Hardy, 1615 NW 14th Avenue, Gainesville, Florida 32611, USA. Over 50 species.

Bird Songs in the Dominican Republic by G. B. Renard. 1981. Two 30 cm discs, TR 520, 594. Cornell Laboratory of Ornithology (as above). About 100 species.

Bird Songs and Calls from South-East Peru by B. and L. Coffey. 1981. One cassette. Produced for Explorers' Inn and the Tombopata Nature Reserve by B. B. and L. C. Coffey, 672 N Belvedere, Memphis, Tennessee 38107, USA. Also available in Peru from Peruvian Safaris S.A., Casilla 10088, Lima 100, Peru. Over 70 species.

The Songs of the Pampas by R. Straneck. 1981. 30 cm disc, Phillips 6347.515. Museo Argentino de Ciencias Naturales, Av. Angel Gallardo 470, Casilla de Correo 220, Sucursal 5-1405 Buenos Aires, Argentina. 60 species.

ANTARCTIC REGION   The area south of the Antarctic convergence (roughly 50° south)

Antarctica by E. Mickleburgh. 1971. 30 cm disc, Saydisc SDX 219. Saydisc Specialised Recordings Ltd., The Barton, Inglestone Common, Badminton, Glos. GL9 1BX. 9 bird species.

Index

PICTURE CREDITS